Robot Programming

A Practical Guide to Behavior-Based Robotics

Joseph L. Jones

Robotic Simulator by Daniel Roth

McGraw-Hill

New York   Chicago   San Francisco   Lisbon
London   Madrid   Mexico City   Milan
New Delhi   San Juan   Seoul
Singapore   Sydney   Toronto

The McGraw·Hill Companies

Cataloging-in-Publication Data is on file with the Library of Congress

Copyright © 2004 by The McGraw-Hill Companies, Inc. All rights reserved. Printed in the United States of America. Except as permitted under the United States Copyright Act of 1976, no part of this publication may be reproduced or distributed in any form or by any means, or stored in a data base or retrieval system, without the prior written permission of the publisher.

1 2 3 4 5 6 7 8 9 0 DOC/DOC 0 9 8 7 6 5 4 3

ISBN 0-07-142778-3

The sponsoring editor for this book was Judy Bass and the production supervisor was Pamela A. Pelton. It was set in Melior by Patricia Wallenburg.

Printed and bound by RR Donnelly.

This book is printed on recycled, acid-free paper containing a minimum of 50 percent recycled, de-inked fiber.

McGraw-Hill books are available at special quantity discounts to use as premiums and sales promotions, or for use in corporate training programs. For more information, please write to the Director of Special Sales, McGraw-Hill Professional, Two Penn Plaza, New York, NY 10121-2298. Or contact your local bookstore.

To Sue, Kate, and Emily

Contents

Preface

I got my first taste of robot programming in the early 1980s when I joined the research staff at MIT's Artificial Intelligence Laboratory. My group was trying to solve a classic challenge in robotics called pick-and-place—make a robot pick up an object at one spot and put it down somewhere else. Given an object and a destination all the robot has to do is to figure out the actual arm and gripper motions needed to move the object to the goal—the sort of thing any two-year-old can do. Four of us worked on the problem for about five years.[1]

Other group members worked on the parts of the program that would generate the large-scale motions of the robot arm, motions to move the arm from one region of the workspace to another. My job was to write the software that would enable our robot arm (see **Figure P.1**) to work out how to move the last few inches toward an object and grasp the object. The solution to the overall problem has many constraints: the robot has to grasp the object at a viable spot; the robot must avoid bumping into anything as it moves about; the robot must avoid violating what are called kinematic constraints.[2]

---

[1]See *Handey: a Robot Task Planner* by Tomás Lozano-Pérez, Joseph L. Jones, Patrick A. O'Donnell, and Emmanuel Mazer, MIT Press, 1992

[2]A kinematic constraint is a limitation on how a robot can move. If the robot can say, position joint A only between the angles 0 and 120 degrees, the con-

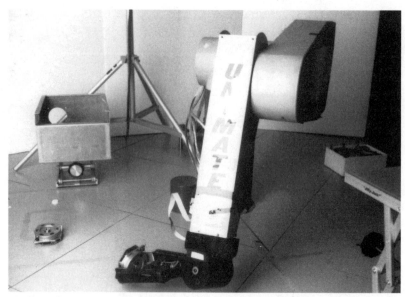

Figure P.1

This Puma model 560 with custom-built gripper was one of the manipulator robots used in the Handey project. In the foreground the robot picks up a motor that will be added to an assembly (contained in the white box) at the back right. The Handey program generates all the joint motion commands needed to move the motor from the point where it is picked up, avoiding all the obstacles, and insert it into the assembly. *(Photograph courtesy of Prof. Tomás Lozano-Pérez of the MIT Artificial Intelligence Laboratory.)*

Yet another part of our work was to write code that would figure out how to reposition the object in the robot's gripper if the initial grasp conflicted with obstacles or the robot's kinematic limits at the putdown point. The tricky bit was that our software was supposed to be completely general—the code had to work for any robot, in any environment, transporting any part.

In order to accomplish all these things we had to first build a world model. A world model tells the robot the geometric shape of every object in the robot's workspace and where every object is located in relation to the robot. And in the same meticulous way that we modeled the environment, we also had to model the robot and to program the equations that described the robot's

trolling program must refrain from instructing the robot to move joint A to 135 degrees.

kinematics—how the robot's joints relate to each other and in which ways and how far each joint is able to move.

Our task was excruciating. Any small error in the world model could cause the robot to collide with an object when the robot tried to execute the motions it had planned. Any little mistake in the equations that describe the robot meant the robot might fail to reach the designated pickup-object or whack something along the way. If one of us accidentally bumped some object in the robot's workspace, thus creating a mismatch between the real world and the robot's world model, the robot would most likely strike that object. And, when at last the robot came up with a successful plan for moving an object from one place to another, the robot's motions invariably looked awkward and unnatural.

We hoped that our work would have the practical result of making manufacturing by robots faster and more flexible. Our software should enable assembly line designers to describe in generic terms what they wanted the robot to do rather than telling the robot in precise detail how to do it. Using the most sophisticated computers and the best thinking available at the time we succeeded in solving an interesting academic problem.

However, our efforts did little to change the way robots manufacture products. It still makes economic sense for assembly line workers to follow the painstaking process of teaching their robots each and every tiny motion needed to perform an assembly. Robots that plan such motions by themselves can't compete—they don't seem to add enough value to earn their keep.

At the same time that I worked on the pick-and-place problem another team at the AI Lab tackled a different robotic challenge—getting autonomous mobile robots to negotiate a real world environment. The Mobile Robot[3] group focused on a different class of robots in a different environment and took an approach fundamentally different from the one my group followed.

Insects fascinated the Mobile Robot folks. They noted that these creatures are a marvel in a minuscule package. In a complex and

---

[3]The Mobile Robot group was established and led by Prof. Rodney Brooks.

dangerous world, insects manage to find food, shelter, and mates. Insects escape from predators. Insects navigate their world and don't get lost. And sometimes insects even seem to cooperate in building large structures and in performing other impressive feats. Yet insects have the tiniest of brains. For many insects, sight is accomplished using primitive vision systems-systems that boast fewer pixels than a cheap video camera. What were dumb bugs doing that put our best robots to shame?

The (partial) answer that the Mobile Robot group and others developed is behavior-based robotics. Behavior-based robotics is having an impact not just in academia but in the larger world as well. Sojourner, the robot that successfully explored a little part of Mars in 1997, used behavior-based programming to achieve its otherworldly feat. But a little floor cleaning robot called Roomba® provides us with a more, shall we say, down-to-earth example of behavior-based robotics. See **Figure P.2**.

Many extol Roomba®[4] Robotic FloorVac as the world's first practical consumer robot. Indeed Roomba® has established a growing

Figure P.2

Roomba® is a widely available floor cleaning robot manufactured by iRobot, Corporation of Burlington, MA. Roomba uses a behavior-based programming scheme. *(Photo courtesy of iRobot Corporation, Burlington, MA)*

---

[4]Many people at iRobot Corporation of Burlington, Massachusetts (www.irobot .com) made crucial contributions to the development of the Roomba® Robotic FloorVac development. The original team included Paul Sandin, Phil Mass, Eliot Mack, Chris Casey, Winston Tao, Jeff Ostezewski, Sara Farragher, and Joe Jones.

presence in a habitat previously far more forbidding to robots than even the dusty plains of Mars—the display shelves at mass-market retailers. Like Sojourner, Roomba® abides by the principles of behavior-based robotics. Those principles endow both robots with significant capabilities: the ability to make do with a small, low-end processor, to respond quickly to sensory inputs; to perform robustly; and to degrade gracefully in the presence of inaccurate data and partial sensory failure.

Because of its modest computational requirements and ease of implementation, behavior-based robotics is very well suited to the needs and abilities of hobbyists, students, and robot enthusiasts alike. By learning the principles of behavior-based robotics you will be able to create robots that are affordable, responsive, robust, fun, and maybe even useful.

Joe Jones

Acknowledgments

Many people helped to bring this book into being. Judy Bass of McGraw-Hill provided the initial impetus by proposing that I write a new book on robotics. The material of Chapter 8 is derived from a project that Ben Wirz and I worked on for several years. Others made important contributions by reviewing the manuscript and offering crucial suggestions. They include: Adam Craft, Matt Cross, Branden Gunn, Danniel Ozick, Paul Sandin, Steve Shamlian, Jennifer Smith, Sue Stewart, Chuck Rosenberg, Clara Vu, Greg White, Bill Wong, and Holly Yanco.

Gratitude is also due to the growing numbers of robotics enthusiasts around the world—hobbyists, students, educators, and researchers alike. These are the individuals who make robots (and books about robots) possible; they comprise the fount from which future progress in robotics will flow.

Introduction

There are many ways to program a mobile robot. The least complex robots have programs written in solder[1] where sensors are connected more or less directly to motors. At the other extreme are robots programmed in high-level languages—languages like Lisp and Java that are often used to support artificial intelligence. But regardless of how the programming is accomplished, all robot programs exhibit some sort of structure or architecture.

Programs composed by beginning roboticists often have a structure that might be described as *ad hoc* or perhaps organic—the programs just grow. Without any overarching principles or methodology, the programmer thinks of a feature and writes the code that implements that feature. A second feature springs from the fertile mind of the programmer and is combined with the first, a third feature comes to be, and so on. As development proceeds, the organic analogy becomes more and more apropos. A bit of code that implements one aspect of the program intertwines with code that implements another, magic numbers take root, and special cases flourish and spread.

---

[1]For an approach to robotics rather different from the microprocessor-based systems in this book you may want to investigate the innovative work of Mark Tilden. Start at the Web site: http://www.nis.lanl.gov/projects/robot/.

The organic approach *does* work. With enough time, patience, and code space, the programmer can coerce an organically structured robot program into producing desired results. But without the guidance of robot-specific principles, such programs become more and more convoluted. It always seems to take longer to implement the next feature than it took to implement the last; programmers often refrain from uprooting obsolete features because bugs tend to crop up when "unused," but highly cross-coupled sections of code are deleted. Even when programmers follow modern tenets of modularity and other good programming practices, robot programs can become cumbersome and fragile.

Robot/Computer Differences

Programming trouble develops because beginning roboticists often miss the importance of a basic fact: robots and computers are different. The goals of a robot program and a computer program are different; the constraints imposed on computers and robots are different—and these differences are crucial. A roboticist must understand and respect the distinctions if he or she is to develop effective robot programs.

Serial versus Parallel

The intuition that programmers develop learning to program computers often does not carry over to programming robots. One aspect of programmer intuition that sometimes fails concerns the issue of serial versus parallel execution. Serial execution is satisfactory for most computer programs, but robot programs demand a parallel approach.

A typical computer program is designed to compute an answer and return a result. At their core, even highly interactive and intensely graphical computer programs like video games follow this basic approach. Computation proceeds in a sequence of steps where, typically, the output of one step becomes the input of the next step. The total time required to reach an answer is the sum of the times taken for each step. Therefore, a faster computer is always a better computer—the user of a faster computer gets the answer more quickly than the user of a slower computer.

Alternately, a faster computer can give a more precise answer or can give more answers in the same time taken by a slow computer. For a video game, computer speed translates directly into better resolution and more realistic simulations.

But computing an answer is *not* the purpose of an autonomous mobile robot. Rather, a robot seeks to achieve a goal or maintain a state while avoiding hazards and traps—very much as a living organism does. The robot must attend *simultaneously* to all of its concerns; e.g., don't collide with anything, don't fall down the steps, don't run out of power far from the charger. Catastrophe might result if, for example, the robot were to neglect watching out for the edge of the stairs while concentrating exclusively on avoiding a collision with the baseboard. A robot crash can be a rather more serious matter than a computer crash.

Plans versus Opportunities

A typical computer program executes a plan—one thing happens after another until a result is reached. But an autonomous robot needs to be opportunistic—sometimes the robot's goal is already achieved; all the robot has to do is notice. For example, suppose we want our robot to find its way to a charger when the batteries are low. A plan-based program deciding to recharge the batteries might first have the robot search for a central beacon in the room, go to the beacon, orient itself in a particular direction (toward the charger), then proceed until the robot encounters the local beacon that indicates the charger.

This sounds like a reasonable plan—unless the robot happens to be positioned only a foot from the charger when the robot decides that the batteries need recharging. In that case, the plan-following robot first moves *away* from the charger to find the central beacon, only to turn back immediately toward the charger. A more sensible approach would have the robot notice that it was next to the charger to begin with, and then act accordingly, rather than follow a plan blindly, ignoring real-world opportunities. Constant input from sensors and a programming paradigm able to make use of the rush of data are needed to enable a robot to take advantage of its opportunities.

Graceful Degradation

Given accurate data and a logically correct program, a computer will return a correct result. Given inaccurate data, the computer's output is unreliable. This observation gives rise to the well-known GIGO adage, "Garbage in, garbage out." The computer depends on a human operator to input accurate data and has little recourse if the data are wrong. Thus incorrect computer responses are ascribed to "human error."

An autonomous robot collects its own data through its sensors. And sensors, as we shall see, often mislead. The less you pay for a sensor, the less inclined it is toward veracity. But even very expensive sensors provide unreliable data in common situations. Because of the unreliability of its inputs, a robot program must be engineered in such a way that the robot works as well as possible under the circumstances. That is, robot performance should degrade gracefully in the presence of inaccurate or missing data. A robot program must not collapse in a heap (as a computer program might) at the first sign of erroneous input.

Behavior-Based Advantages

Behavior-based approaches to robot programming excel at parallel execution, opportunistic goal realization, and graceful degradation. Programming a robot according to behavior-based principles makes the program inherently parallel, enabling the robot to attend simultaneously to all the hazards it may face as well as the serendipitous opportunities it may encounter. Further, behavior-based robots can easily accommodate methods that allow performance to degrade gracefully in the presence of sensor error or failure.

What's All the Fuss?

Experienced programmers may be wondering: Is there anything new here? Isn't an autonomous mobile robot really just an example of an embedded system plus a real-time operating system? Might "behavior-based programming" be nothing more than a fancy name applied to commonplace practices?

Mobile robots do indeed qualify as embedded systems. And whether the robot's software runs under a commercially supplied real-time kernel or is implemented in raw code, every robot must have some semblance of a real-time operating system. Further, the good practices a roboticist follows in constructing a behavior-based program are not inconsistent with the good practices an embedded system programmer might follow writing code that controls, say, a DVD player or a cell phone.

However, by themselves, common practices for writing embedded system code are not sufficient for constructing effective robot programs. Autonomous robots, as will become increasingly clear, regularly confront challenges rarely faced by other embedded systems. Dealing with these challenges calls for the additional set of organizing principles that behavior-based programming supplies.

Focus

The behavior-based robot programming paradigm is an eminently practical one and likewise, throughout this text, my treatment of behavior-based robotics will focus more on practical issues than on academic rigor.[2] My intention is to provide you with important fundamentals that will help you program robots effectively. I hope that you will take away from your study of behavior-based robotics an appreciation of the power and scope of this robot programming method and that you will feel confident implementing a behavior-based approach in your future robotic projects.

In this text, I will explain robot programming theory, offer examples of successful robot programs, and relate insights that I have found useful. But ultimately you cannot learn robot programming from a book. To learn robot programming, you must program robots. Unfortunately, because of cost and other complications, not everyone has immediate access to a programmable

---

[2]For an excellent example of a more rigorous approach see *Behavior-Based Robotics* by Ronald Arkin, MIT Press, 1998. See also course notes prepared by long-time behavior-based robotics researchers Ian Horswill at http://www.cs.northwestern.edu/academics/courses/special_topics/395-robotics/.

mobile robot. To address this difficulty, my colleague, Daniel Roth, has developed a robotic simulator, BSim; we have integrated BSim with the text. To access BSim please visit the Web site: www.behaviorbasedprogramming.com.

We recommend that you make a practice of running BSim frequently as you work toward understanding the concepts presented here. As a pedagogical tool, robot simulators can be quite helpful. BSim lets you isolate aspects of robot behavior and slow down processes so that you can easily observe what is going on. However, as a predictor of how a physical robot will interact with physical objects, simulators are notoriously unreliable. *Do not assume that what works in simulation will work in the real world.* To fully appreciate how a robot behaves in the physical world there is no substitute for programming an actual, physical robot.[3–6]

Prerequisites

What do you need to know before you begin? Because robots have such a broad appeal and attract a diverse set of enthusiasts, it is hard to choose a firm set of prerequisite knowledge. Each reader will bring a different set of interests, experience, and willingness to acquire missing understanding. However, in general,

---

[3]For a compilation of inexpensive robots see *Personal Robotics: Real Robots to Construct, Program, and Explore the World* by Richard Raucci; AK Peters, Ltd., 1999. More recent offerings may be found by searching the Web for "Robot Kits." Also, try http://www.robotstore.com.

[4]The possibilities for experimenting with behavior-based robotics have expanded in recent years. LEGO offers a product called Mindstorms (http://www.legomindstorms.com) powered by the RCX, a user-programmable controller. The RCX, though it possesses only a limited number of inputs and outputs, facilitates the construction of many interesting robots. Enthusiasts have greatly extended the abilities of the RCX by creating new software systems (some of them free) compatible with the controller. One such system is called LeJOS. This system allows you to program the RCX in Java. LeJOS even provides libraries that promote a behavior-based approach! Look for information on LeJOS at: http://lejos.sourceforge.net/.

[5]A respected organization that promotes learning about robots and behavior-based programming is the KISS Institute for Practical Robotics; see http://www.kipr.org/.

[6]For the authoritative text on building robots using the LEGO construction system, see *Robotic Explorations*, Fred Martin, Prentice-Hall, Inc., 2001.

a reader will have an easier time if he or she has a working knowledge of algebra and trigonometry and has some experience with vectors. You will miss little if you have yet to master linear algebra and calculus. It will be helpful to have basic familiarity with computer programming before reading this book, but no great depth of experience is assumed.

Organization

In Chapter 1, we use BSim to observe the behavior of a working system, a simulated robot. The simulated robot exhibits interesting and complex interactions with its environment. To understand these interactions we take a step back and ask: what exactly is a robot and what are its essential components?

In Chapter 2, we review the feedback control system, a traditional method of connecting sensing to actuation. Returning to BSim, we observe the functioning of elementary control systems. But from the very beginning we will learn the painful lessons of how good control systems can go bad.

In Chapter 3, we begin to build primitive behaviors, learning about triggers and about ballistic versus servo behaviors.

Chapter 4 deals with arbiters, the software construct that all behavior-based systems must have to manage conflicts between behaviors.

Putting together all that we have learned, in Chapter 5, we can write complete programs. We commence filling a bag of useful robot programming tricks.

With the basic science now mastered, we proceed to the art of robot programming in Chapter 6. Here we address how a problem statement can be converted into a robot program. Principles and heuristics to guide this perilous but necessary step are suggested.

In Chapter 7, we take a software-centered look at some common sensors, how they function, and the ways that sensor output can be misleading.

Chapter 8 recounts a case study of a behavior-based robot implementation. This implementation can be instantiated[7] using an 8-bit microcontroller using the C computer language.

Robotics is a vital and growing field and in Chapter 9, we take a brief look ahead. A few of the sorts of robots that may soon be possible are discussed, as are some of the key missing elements holding back the arrival of these robots.

Programming a robot means, among other things, specifying how a robot moves. Appendix A provides some of the frequently used details that a program must incorporate to control robot motion.

Appendix B describes BSim, its functioning, and construction. BSim is open-source software written in Java; you may wish to extend or specialize the program for your own purposes. Consult the BSim Web site for details.

In Appendix C are a number of simple functions that I have found beneficial in the construction of many different robot programs.

Pseudocode is used throughout the text to describe the workings of robot programs. Appendix D describes the informal formalism used in the book.

In my experience, developing robots requires two things above all else: technical proficiency and unfettered creativity. The exercises at the end of each chapter are intended to promote both. Many of the exercises seek to expand your proficiency by asking you to compute a value or solve an equation. Many others challenge you to think about something. The latter exercises may have no answer or many answers—their purpose is to stimulate creative thought.

Building and programming a robot is very much an act of creation—an undertaking that I continue to find immensely satisfying. I hope you will enjoy your endeavors in the field of robotics as much as I have mine.

---

[7]To instantiate something is to build a concrete example of an abstract concept.

1
Autonomous Mobile Robots

To begin our study of behavior-based robotics, let's skip ahead to one of the fun parts. We will use BSim to demonstrate a robot executing a behavior-based program. Please visit the Web site www.behaviorbasedprogramming.com. There you will find a description of how to use BSim (a sample screen is shown in **Figure 1.1**). Select the Collection task, start the simulator, and watch as the simulation runs.

Example: Collection Task

How would you describe the actions of the simulated robot? From a top-level perspective, you could say that the robot is searching for a certain type of object (pucks). When the robot finds a puck, the robot then pushes the puck to the vicinity of the light source. The robot drops the puck off near the light source and then goes to look for another puck. If you are being especially precise, you might add to your description that while the robot finds and delivers pucks, the robot simultaneously avoids or escapes from encounters with other objects.

The sort of multifaceted, overall activity that the robot exhibits in this example we shall call a *task*.[1] A task is the

---

[1]A task is to a robot as an application is to a computer.

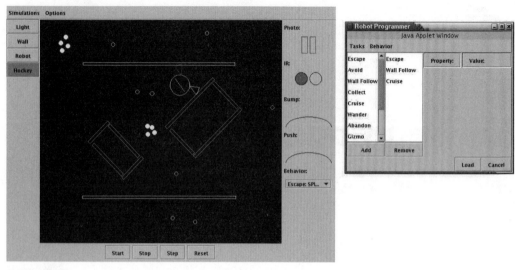

Figure 1.1

The BSim robot simulator enables users to build behavior-based programs for a virtual robot and view their operation in a simulate world. Here the robot executes a sample task. The simulated world is visible in the dark central area. Narrow rectangles represent walls, small open circles are pushable pucks, and closed circles represent light sources. The large circle is the robot. Wedge-shaped areas emanating from the robot show the field of view of short-range obstacle sensors. The status of the robot's sensors and operating program line the right side of the window. The smaller window on the right enables the user to program the robot.

kind of job you might ask another person to do. Tasks often have rather simple descriptions. If you wanted someone to perform the robot's job in this example you might say, "Please move all the pucks next to the light." But, as we shall later see, encoded in that deceptively simple phrase is a world of intricacy and subtlety.

Our ultimate purpose is to learn to design and program autonomous mobile robots. Starting with the description of a useful task that we wish the robot to perform, we must specify the physical features and write the appropriate software that will enable a robot to accomplish that task. One of the most powerful weapons we have for attacking complex problems of this sort is reductionism. Reductionism means that we try to reduce or decompose a large, difficult problem into a set of smaller, simpler problems that we can more easily understand. But what principles should we use to chop a big robot problem into smaller, more manageable ones? For autonomous mobile robots, behavior-based approaches have proven effective.

Taking a behavior-based approach means that we try to devise a set of simple behaviors (algorithms connecting sensing to actuation) that, *when acting together*, produce the overall activity we desire. The Collection task is implemented in exactly this way. The number and nature of the simple (we will call them *primitive*) behaviors that implement the Collection task probably are not obvious to you. However, before you read further, please observe the simulation for a time while considering this issue. You will likely begin to form opinions about how the Collection task is accomplished; see if you can ferret out the elements, the primitive behaviors, that make it work.

The major components of the simulated robot are shown in **Figure 1.2**. The bumper can determine if a collision has occurred and can sense whether the contacted item is sliding (like a puck) or is fixed (like a wall). BSim is implemented in such a way that the light sensors can see over the pucks, but walls block the light. The walls and pucks reflect infrared radiation (IR), so that

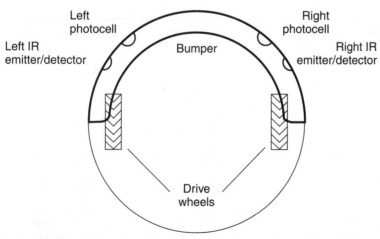

Figure 1.2

The BSim robot includes a simulated bumper that detects collisions with objects, IR proximity detectors that attempt to detect objects before a collision occurs, and dual photocells that can determine the direction and intensity of a light source. Robot motion is enabled using two independently controllable drive wheels visible through the robot shell.

the robot can detect them with its IR sensors. The light sources hang from the ceiling so that the robot can pass under them.

Figure 1.3 shows the *behavior diagram* of the program that implements the Collection task. A behavior diagram is a graphical tool that helps us understand what the robot program does and how it works. Most details in the figure are unimportant at this stage, but do take note of the major divisions. The robot can be thought of as having three parts devoted to sensing, actuation, and intelligence. The intelligence section takes input from the robot's sensors and sends its output to the robot's actuators. Internally, the intelligence section is composed of several primitive behaviors and an arbiter. It is the nature of these primitive behaviors that you have been trying to identify.

The Collection task is composed of only six primitive behaviors. They are called Escape, Dark-push, Anti-moth, Avoid, Home,

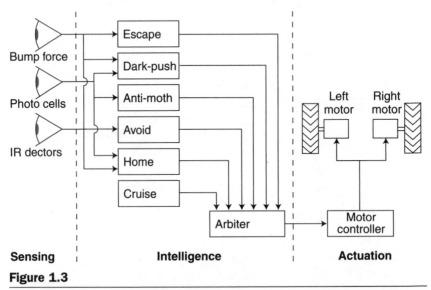

Figure 1.3

A behavior diagram graphically represents the operation of a behavior-based robot. At the highest level, the robot is composed of elements that provide sensing, actuation, and intelligence functions. Information about the robot's environment flows in through the robot's sensors to several primitive behaviors in the intelligence section. Behaviors named Escape, Dark-push, Anti-moth, Avoid, Home, and Cruise implement the Collection task by computing motion commands for the robot. A structure called an arbiter combines or selects the commands and sends a final command on to the robot's motors.

and Cruise. If the robot collides with an object that the robot cannot push, the Escape behavior directs the robot to back up and spin, thereby choosing a new direction. Dark-push tries to avoid pushing pucks when the light is not visible to the robot; this is the case when the robot faces away from the light. Refusing to push pucks in the dark usually prevents the robot from pushing pucks away from the light. The Anti-moth behavior causes the robot to turn away when it gets too close to the light. This lets the robot drop off a puck in the vicinity of the light. The Avoid behavior monitors the robot's IR proximity detectors. Normally an avoid behavior would cause the robot to turn away from obstacles, but the Collection task configures Avoid in the opposite way. When the robot senses an object, the robot turns *toward* the object. This helps the robot find pucks. If the object happens not to be pushable, the Escape behavior takes over to drive the robot away from the obstacle. The Home behavior allows the robot to orient itself toward the light source and move in the direction of the light. The Cruise behavior has the robot drive straight when the robot otherwise doesn't know what to do.

Perhaps some behaviors that you expected to find are missing from the roster. There is, for example, no explicit Find Puck behavior, no Push Puck behavior, and no Drop Puck behavior. Although the robot seems to perform these three activities purposefully, in actuality the Push, Find, and Drop operations *emerge* from the interaction of more primitive behaviors. This theme of complex behaviors emerging from simpler ones is a central feature of behavior-based robotics.

Watching the simulator, you may have noticed that the Collection task does not always proceed smoothly. Sometimes the robot drops a puck on the way to the light only to come back and pick up the puck later. The behavior of the system is thus not purely deterministic,[2] but rather contains a random component. You never know exactly how the robot is going to get the pucks to the light. But at the same time, the robot's overall

---

[2]In a deterministic program, events follow each other in a completely predictable way. Given a particular initial configuration and a particular sequence of sensor readings, the robot always responds in exactly the same way.

behavior is very robust—no matter the bumps and missteps along the way, ultimately the robot collects the pucks.

These are basic characteristics of behavior-based systems. But before we plunge into a sea of details, let's step back to view the big picture and acquire a better grounding in the fundamentals.

Robot Defined

There are many definitions of the word robot, but I think of a robot[3] simply as: *A device that connects sensing to actuation in an intelligent way.* This minimalist characterization takes for granted a couple of assumptions. First, the thing the robot senses is the external world, not just the robot's own internal state. Second, at least some of the actuation is applied to the robot itself in such a way that the robot moves (see **Figure 1.4**). (Without these assumptions, a sedentary lawn watering system, whose only sensor is a timer and whose only actuator is a servo valve on a water line, would qualify as a robot. Most unsatisfying!) Making intelligence part of the definition of robot seems to

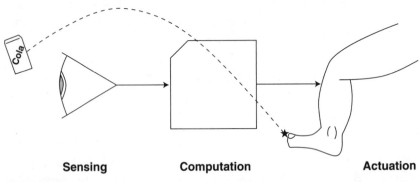

| Sensing | Computation | Actuation |

Figure 1.4

In its simplest form, a robot senses the world, performs some computations on what it has sensed, and then acts. The robot's actions can affect both the robot and the world.

[3]Teleoperated mobile devices, platforms controlled exclusively by a remote human operator, are popularly referred to as robots. However, for purposes of this text, a robot must be autonomous to qualify as a robot. Thus, in general, robots mentioned here will be assumed to be autonomous.

add a subjective component. But all that I mean by intelligence in this context is that the robot processes sensor information into actuator outputs with some minimum level of complexity.[4]

A robot is not a computer. Computer scientists agree on the abstract definition of a computer. And although one computer may be faster or slower than another or may have more memory than another, in a fundamental sense, all computers are equivalent.[5] The same high-level program that solves a problem on a multimillion-dollar supercomputer can, in principle, solve the problem on an 8-bit microprocessor—eventually.

The equivalence of all computers makes it far easier for computer scientists to analyze computer programs than it is for them to analyze robot programs. Using the notion of equivalence and the definition of computability, computer scientists can prove theorems about computation. Often scientists can decide in advance if a computer is able to solve a particular sort of problem and, if so, how long that solution might take.

The practical effect of the equivalence of computers is that when you sit down to write a program in a high-level language, you need give almost no thought to which computer will run your program. Is the hard drive interfaced to the computer via a FireWire or a USB connection? You don't care. Will the computer be powered by 120-volt AC and run in the United States or will the computer be plugged into 240-volt supply in a European country? It makes no difference. Will the computer use an optical mouse or a mechanical mouse? Such issues simply have no relevance as you compose your program.

---

[4]People disagree about what properly constitutes a robot; my definition does not resolve those disputes.

[5]Remarkably, the equivalence of all computers was proven in 1936—a time when the first electronic computer was still a decade away. Alan Turing, a British mathematician, is responsible for the proof; he invented a concept called the universal Turing machine. All computers are Turing machines and because of this, all computers share certain fundamental properties. See *Computer Power and Human Reason* by Joseph Weizenbaum; MIT Press, 1976 for a readable explanation of Turing machines. Or review Turing's original paper, "On Computable Numbers, with an Application to the Entscheidungsproblem," Alan Turing, *Proc. London Math. Soc.* Ser. 2–42 (Nov. 17, 1936), pp. 230–265.

But the logic that proves the equivalence of computers does not hold for robots. In general, robots are decidedly *not* equivalent to each other. Rarely can one devise a theorem proving that a robot will be able to perform a particular task. Different robots have different sets of sensors and different sets of actuators. Robots come in different shapes and have sensors positioned in different places. In contrast to computers, such differences are not academic—when it comes to robots, these differences are critical. Two robots constructed or equipped differently but designed to accomplish the same task may require not just different code, but different approaches to the problem. Even robots that are identical except for where sensors are positioned on the robot may require radically different programs. With robots, we don't get to ignore the details.

Still, we can make some statements that hold true for all robots. Here are a few: Robots have sensors that measure aspects of the external world. Robots have actuators that can act on the robot and on the world. The output of a robot's sensors always includes noise and other errors. The commands given to a mobile robot's actuators are never executed faithfully. Not all these statements are happy ones, but buck up! It's the challenge that makes robotics interesting.

Let's consider a robot's components in a bit more detail.

Sensing

To act effectively in an unknown and changeable world, a robot must collect information about the world. Sensors provide that information. A sensor is a transducer that converts a physical quantity into an electrical signal. Once this signal is digitized, the robot's computing elements can use the signal to reason about the external world.

I find it helpful to think of sensor-provided signals as answers to questions that the robot asks about the external world. Perhaps hundreds of times per second a robot asks questions like: "Is the left bump switch triggered?" "Does the right-pointing IR sensor detect a reflection?" "What is the numeric difference of the signals from the left and right photocells?" Together, answers to questions such as these tell the robot its current situation; knowing the situation, the robot's program can then choose an appropriate response. When sensors answer the robot's questions correctly and unambiguously, the robot accomplishes its task. But if the robot can't get good answers or, worse, if it asks poor questions, success becomes elusive.

Take another look at the Collection task running on BSim. BSim's sensor window shows the instantaneous values of the robot's sensors. Observe in particular the display of the photocells while the robot is homing on the light source. (See the online help information if you have trouble identifying the correct display.) The left and right photocells point diagonally forward. The left photocell points more directly toward the light than does the right photocell when the robot points to the right of the light. In this configuration, because of the more direct pointing, the left photocell detects more light than the right. When the robot points to the left of the light source, the right photocell sees more light than the left. The robot's program uses this difference to move the robot in such a way as to equalize the light intensity measured by each photocell and thus home on the collection point. And this is where the conceptual trouble with sensors begins.

In our example, the question we want to answer is, "Which way is the collection point?" But "collection point" is a human concept. The robot knows nothing about such things. All the robot has to work with are numbers—the digitized electrical signals produced by sensors. In our example, we have arranged things so that when the robot points toward what we think of as the collection point, the photocell light level difference is zero. Thus the only question the robot needs to ask (or is able to ask) is, "What is the difference between the left and right photocell readings?" The relationship between the human concepts with which we reason and the numbers the robot computes is in the programmer's mind, not the robot's mind.

Our first duty as robot programmers is to make sure this relationship, between the questions we would like the robot to ask and the ones it is able to ask, is as sound as possible. We must always consider such things as: Under what circumstances does the difference between the photocell readings *not* indicate the direction to the collection point? Could there be another light source that will fool the robot? Do the two photocells report the same numeric value when exposed to the same intensity of light? And so on. Being always mindful of the shortcomings of our sensors and our concepts will help us program more effectively.

Actuation

Having decided what to do, the robot makes it so by sending commands to the actuators. Actuators are transducers that perform an operation inverse to that of sensors—an actuator converts an electrical signal into a physical quantity. Actuators can include devices like speakers that convert electricity to sound. Sometimes, even LEDs that convert electricity to light are thought of as actuators. But in the context of mobile robots, an actuator usually consists of one or more electric motors connected via a gear train to a wheel, leg, arm, or gripper.

The virtual robot in BSim has two actuators, the left and right wheel motors. Actuators are less philosophically challenging than sensors, as they raise fewer issues of concept or perception. Send an electric current to an actuator and the actuator does something. Still, like sensors, actuators have their own ways of confounding us when we try to use actuators to maneuver a robot. In particular, real-world actuators don't always do as they are told. Send, say, 500 milliamperes of current to both drive motors for one second and, in one situation, the robot will move one foot forward. But send exactly the same current for the same time to the motors when the robot is in some other circumstance (for example, straddling the divide between a tile floor and a shag carpet) and the robot will translate half a foot forward while spinning 30 degrees to the left.

Intelligence

The robot's design, the robot's program, and (some would say) the robot's environment combine to produce the robot's intelligence. But a robot cannot behave in a particularly intelligent way if it ignores the difficult issues surrounding sensors and actuators. Put bluntly, a robot program that does not accommodate the shortcomings of sensors and actuators is a program that will fail. Indeed, it was sensor and actuator issues coupled with even greater challenges arising from the unknown and changeable operating environments that spurred the creation of behavior-based robotics.

Mobile and Immobile Robots

Mobile robots face unique challenges not shared by computers (see **Figure 1.5**) or even by manipulator-type robots. It is these challenges that make programming mobile robots both interesting and difficult.

Figure 1.5

In psychological terms, computers are delusional. Computers formulate their plans in a made-up virtual world—a world that need have no connection to physical reality. It is relatively easy for a robot to formulate optimal plans in the sort of idealized world modeled here. The staggeringly difficult part is ensuring that, at all times, the model matches every significant aspect of the real world.

Mobile robots and manipulator robots typically operate in strikingly different environments. Specifically, the sort of manipulator robots used in industry inhabit environments that are highly structured and relatively static (**Figure 1.6**); mobile robots that perform real-world tasks occupy largely dynamic and unstructured environments (**Figure 1.7**).

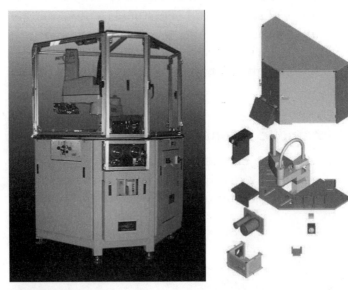

Figure 1.6

A typical robotic work cell is structured and predictable. The robot programmer accounts for every object in the cell. No object is permitted to change its position or to enter or leave the cell without the robot's knowledge. As shown in an exploded view on the right, a solid model of the robot's environment can be constructed. Such a model can be used to automatically plan robot motions. *(Photo of Industrial Robotic cell for small parts processing in the electronics industry, courtesy of RAPID4mation Inc., Portland, Oregon.)*

A highly structured environment is one where the placement of objects is deliberate. The nature and location of each object is known. Non-rigid objects, for example cloth or plants, are typically excluded from structured environments because it is difficult to describe to a robot the shape and motion of such objects in a simple mathematical way. Robotic work cells are highly structured, office buildings and private homes are less so, while forests and swamps are very much unstructured.

Figure 1.7

This six-legged mobile robot, called Ariel, functions in an outdoor environment. The robot must respond quickly to changes and unexpected events. *(Photo courtesty of iRobot Corporation, Burlington, Massachusetts.)*

A dynamic environment is an environment where things change. Objects in a dynamic environment may alter their position or orientation independently of the robot. A work-cell robot can accept some dynamic elements (conveyer belts transporting parts, for example) as long as the robot is able to predict or measure the position of those elements. But a mobile robot usually has no *a priori* way of knowing what objects are present in its environment or when or how they may move.

There is one more important difference between manipulator and mobile robots. Manipulators always know where they are, but mobile robots rarely do. Manipulator robots are made up of some number of links; a joint connects each pair of links. This can be seen in **Figure 1.8**. Each joint includes an encoder that measures the angle[6] between the adjacent links to a high degree

---

[6]These comments assume revolute joints, but apply equally well to manipulator robots with prismatic joints. See *Robot Manipulators: Mathematics, Programming, and Control,* by R. P. Paul, MIT Press, 1982.

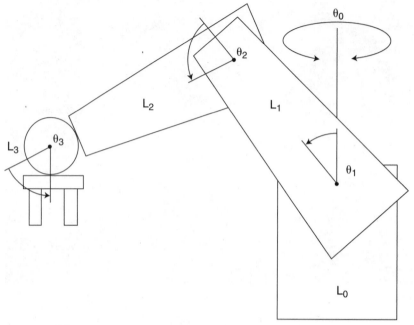

Figure 1.8

A manipulator robot is constructed from links and joints. A servomotor and encoder associated with each joint, j, establishes with great accuracy the angle θ_j between adjacent links L_j and L_{j-1}. Knowing the exact value of each joint angle enables the manipulator robot to compute the precise position of its end effecter (gripper).

of accuracy. When given a command to move the end point of the manipulator to a particular (x,y,z) position, the end point will move to almost exactly that position. The encoders in the joints tell the manipulator whether it has reached the correct position—the robot adjusts the current to its servomotors until the joint moves to the right angle.

Mobile robots often have encoders in their wheels to measure the robot's movements. But there is a big difference between wheels and joints: wheels can slip. Yet slip is not the most significant problem—the environment is. Hard "flat" floors are never flat and carpets have a nap that preferentially nudges the robot in one arbitrary direction.[7] Thus, commanding a mobile

---

[7]Try the experiment by stepping onto a medium pile carpet (the newer the better). Careful observation will reveal that your foot actually moves to one side a bit with each step you take as the carpet fibers fold over.

robot to move to a given position is chancy. A mobile robot tries to execute a positioning command faithfully by monitoring the motion of its wheels just as a manipulator monitors its joint encoders. But small hills and valleys and nap-nudges in the environment constantly lead the robot astray. Watching only the motion of its own wheels, the robot has no way of knowing whether it has arrived at the place it was told to go. (See Figure A.8 in Appendix A.)

Manipulator robots got their commercial start in the 1950s when Joseph Engelberger founded a company called Unimation. Unimation's Unimate robot began earning its keep at a General Motors plant in 1961.[8] Why is it that compared to their sedentary cousins, mobile robots have taken such a long time even to start to become useful? The difficulty of finding ways to cope with unpredictable environments and uncertain positioning has held mobile robots back.

But cope with these problems (and many others) we must if we hope to put mobile robots to work. We must discard the seductive but illusory world inside a computer and the sterile artificial world of the work cell and instead accept the challenge of dealing with the messy and frustrating issues of the real world.

Responding to the Challenge

Mobile robots don't know what they will find in their environment (the environment is unstructured); the position of nearby objects can change without warning (the environment is dynamic); and mobile robots usually don't know where they are (position uncertainty grows as the robot moves). Despite these fundamental issues, until the mid-1980s most researchers tried to force mobile robots into the same mold as manipulator robots. That is, researchers tried to account for every object, build a world model, plan a series of motions, and then execute the plan. But that approach didn't work very well. The amount of computation involved forced mobile robots to move and

---

[8]See *Mind Children* by Hans Moravec. Harvard University Press, 1998.

respond very slowly, and their slow response made them vulnerable to unanticipated changes in their environment.[9] Often, by the time a planning robot got around to executing its plan, the dynamic environment had changed to the point that the plan was rendered obsolete.

A new approach was needed.[10] Rather than relying on an omniscient programmer to tell the robot where everything is, let the robot find out for itself. That is, give the robot sensors that can detect objects in the robot's environment. Because objects can move, don't just sense the objects once and remember where they were, but instead sense the objects continuously and react immediately when motion occurs. If you want to make progress in robotics even though robots don't usually know where they are, then tackle problems that don't require knowledge of absolute position.[11] And most important, build a robot control system designed to handle unstructured, dynamic environments. This is exactly what behavior-based robotics attempts to do.

Robot's World View

To program a robot effectively, we must see the world as the robot sees it. This requires that we largely abandon the familiar

---

[9]Researchers weren't being dense in the days before behavior-based robotics when, as my Handey-project group did, they insisted on using world (and robot) models; they were trying to be rigorous. It *is* possible to analyze robots in a general way, analogous to the way that computers are analyzed *if* we use the formalism of a world model. Unfortunately, the baggage that world models force robots to carry is so cumbersome that in real-world environments, robots collapse under the strain.

[10]In some ways the "new approach" is a case of back to the future. In the late 1940s and early 1950s, a Kansas City-born, British neurological researcher, W. Grey Walter, built some autonomous robots that rivaled the best of the early artificial intelligence robots. Walter's robots, Elmer and Elsie, could wander about without getting stuck, could find their way into their hutches by following a light, and could recharge themselves. Each robot had a brain consisting of two vacuum tubes. Some consider Walter's creations to be the world's first behavior-based robots. You'll find more details on the Web and in Walter's fascinating book *The Living Brain*, W. W. Norton & Company, Inc., 1953.

[11]It should be noted that behavior-based robots have no trouble incorporating absolute position information and using it effectively. Rather, it is the case that such information is rarely available in real-world situations and when it is, the cost is typically high and/or the resolution is low.

views and concepts we use to understand events and to communicate with others. Looking around a room, any person can easily identify chairs, tables, and other people. We know where the walls are, we recognize places we have been before, and we can reason from a rich set of concepts. People have mechanisms, evolved over millions of years, that enable us to interpret the chaotic sensory information we take in. People have hardwired facilities for noticing subtle motions, recognizing faces, and processing sounds. We understand the person speaking to us even when we are surrounded by a dozen other conversations, all taking place at the same time.

Robots start from zero. A robot with a sophisticated sonar sensor does not sense, say, a chair leg, a wastebasket, or an umbrella stand. As indicated in **Figure 1.9**, no matter how complex and interesting its environment, the robot's view of the world col-

Figure 1.9

Humans experiencing the world map sensory information into a rich set of high-level concepts such as friend, rain, and running. Oblivious to all this, a robot viewing the same scene with a sonar sensor experiences the world as a single number—the time of flight of a sonar pulse. *(Drawing courtesy of Sara Farragher.)*

lapses into a single number—the time between the transmission of a sonar ping and reception of an echo. A robot equipped with a pyroelectric sensor does not recognize that a human has crossed its path; rather, the robot notices only that the voltage output by the sensor has just changed polarity.

Programming a robot means accepting this impoverished view of the world and taking effective action—it *can* be done but we have to change our perspective. People have the complex world-view that we do because this view contributes to our survival. Insects have a much simpler view and yet, given their environment and goals, manage quite handily. Insects, for example, don't need to identify chairs or tables or talk about furniture with other insects; they can, however, discover a food source on a table or in the crevice of a chair and carry nourishment back to the nest. The key to effective performance for a human, insect, or robot is matching sensory, motor, and intellectual abilities with the tasks that each must perform.

We do not need to duplicate human-level performance to build a useful robot any more than, say, a grasshopper needs a high school diploma to perform the duties of a grasshopper. And a floor-cleaning robot need not understand that its bumper is, for example, contacting a leather-upholstered sofa imported from Milan. All this robot needs to know is that its bumper is pressed and therefore, a rotate-in-place command must be sent to the motors.

Sensing is the bread and butter of behavior-based robots. The less structured and more changeable the robot's world, the more dependent the robot is on its sensors. But not just any sensing will do. The particular sensing requirements depend critically on the task and on the environment. Once the designer has a thorough understanding of the task and environment, he or she can proceed to select the sensors, actuators, and algorithms best able to accomplish the task in the given environment. There is no safe extreme—failing to supply the robot with a crucial sensor can lead to failure; incorporating a sensor that does not contribute to the task may lead to economic failure.

What does a behavior-based robot do with the sensory information it collects? The robot connects that information as directly as possible to actuation. Thus behavior-based robots are typically highly *reflexive*. As soon as a relevant condition is recognized, the robot takes appropriate action. Unlike the manipulator robot described in the Introduction, a behavior-based robot does not first collect all information (relevant or not) about its environment, plan what to do, and then execute the plan. Instead, as soon as a behavior-based robot has relevant information, it acts on that data.

All these features are apparent in BSim's Collection task. The robot does not plan—it reacts. But the robot's reflexes, and the way they are combined, are carefully engineered to elicit the overall behavior we want. When the robot is not pushing a puck, the robot wanders about. When the robot bumps into a puck, the robot responds by homing on the light. If the robot bumps into an obstacle that it cannot push, the robot reacts by avoiding that object. When we put all these simple actions and responses together, the behavior that we desire emerges: the pucks are moved to the vicinity of the light.

An occasional misstep along the way to accomplishing a goal is a common characteristic of behavior-based systems. A deterministic plan-based system would not abandon pucks far from the light only to return for them later. Such a system would likely never move pucks in the wrong direction. But plan-based systems are brittle[12] and often simply fail to work—an assumption is violated or the world changes during execution and the system stops, unable to proceed. Behavior-based systems, by contrast, strike a reasonable bargain—they trade away brittle determinism acquiring in its place robust chaos.[13] As long as a system makes positive progress more often than it

---

[12]When computer scientists talk about brittle performance, they mean that small errors on the input side of a program cause large changes or catastrophic failure on the output side. By contrast, a robust program is one that suffers only a small performance decline when small input errors occur.

[13]The slogan of the MIT AI Lab's Mobile Robot group was, "Fast, cheap, and out of control!"

makes negative progress, the system will ultimately accomplish its goal.[14]

Summary

In this chapter we have taken a quick look at the raw material we will use for robot programming—sensors, actuators, and reflexive behaviors. We've thought abstractly about how the environment affects sensors, how sensed information is processed, how actuators deal with that output, and how the environment can affect actuation. The lessons we've learned include:

- An autonomous robot is a device that connects sensing to actuation in an intelligent way.

- Robots accomplish tasks.

- To perform its task, a robot must ask certain questions about its environment and situation.

- Sensors provide answers to the questions a robot must ask.

- Unambiguous answers are necessary for effective performance.

- Sensors are rarely able to answer the exact question we would like the robot to ask. We must tailor our methods to the questions sensors are able to answer.

- The details of the sensors used and their placement on the robot critically affects the way the robot operates.

- Complex global behavior can result from simple behaviors acting together.

---

[14]And how does the robot know when the goal is accomplished? In our example, it doesn't. Unless we can provide the robot with a no-pucks-remain-to-be-found sensor, the Collection task will run indefinitely, always on the lookout for a wayward puck. Relying on only local information, behavior-based methods exhibit robust performance. But inferring global truths from only local information is challenging. This fact is sometimes seen as a deficiency of behavior-based methods. Consequently, melding the advantages of behavior-based programming with more deterministic sorts of robot control is an active area of research.

- The insect-intelligence paradigm provides us with a powerful method for thinking about and programming robots.

Exercises

Exercise 1.1   Two manipulator robots that operate in the same work cell have partially overlapping workspaces. Would you describe this system as more structured or unstructured, dynamic or static?

Exercise 1.2   You wish to robotize a cog railroad that carries passengers to the top of a mountain popular with tourists. How would you characterize the environment of the cog railroad robot?

Exercise 1.3   What sensors would you use to implement the cog railroad robot of the previous exercise? Would you install sensors on the robot locomotive, along the track, or both?

Exercise 1.4   If a governmental safety agency reviews your design for the passenger-carrying cog railroad of Exercise 1.2, will the design be approved? Why not?

Exercise 1.5   Suppose you have a circular mobile robot equipped with 16 sonar sensors arranged in a ring around the perimeter of the robot. Each sonar sensor returns a number that (ideally) represents the distance from the sonar to the nearest solid object. Describe how you would use the range information from the sonar sensors to identify an open door that the robot finds while following a corridor—what pattern of sonar reading would such a feature create?

Exercise 1.6   Describe three electromechanical systems that, according the definition of robot stated in this chapter, do not qualify as robots. Name one device, not commonly thought of as a robot, that does qualify.

Exercise 1.7   Besides robotic work cells, what other places could be considered highly structured environments?

Exercise 1.8   Put these environments in order of increasing structure: airport runway, cave, office building roof, dirt road, cornfield, and ski slope. Place the same environments in order of decreasing changeability. Which environment would you expect to be least challenging for robot operations?

Exercise 1.9 Members of the general public often assume that the key to better-performing robots is simply faster computers with greater information storage capacity. Is this assumption correct?

Exercise 1.10 In the animal world, brain size (or perhaps the brain/body radio) roughly correlates with computational power—the bigger the brain, the more intelligent the creature. But is it the case that species success correlates with brain size? Suppose we could transplant the brain of, say, a mouse into a mosquito (without changing the mosquito's sensors, actuators, or basic behavioral pattern). Would having the brainpower (and corresponding weight and volume) of a mouse make the mosquito more successful? Were the mosquito large enough to accommodate a mouse brain, could it function as a mosquito at all? What would the mosquito lose by acquiring a bigger brain?

Exercise 1.11 List several sorts of tasks that a robot might be able to perform. What characteristics make a task well suited to autonomous execution?

Exercise 1.12 List several useful tasks that do not seem amenable to robotic execution at the present time. What makes these tasks ill suited for robots?

Exercise 1.13 Suppose that you have a land-mine detecting sensor. The sensor has a 95 percent probability of identifying a buried mine that is within one foot of the sensor. At 10 percent of tested spots, the sensor reports that a mine is present when there is no mine. Would this sensor be useful for building a mine-removing robot? What issues would you face and what else do you need to know to design such a robot? Does cost of the sensor or the robot matter? When the robot finishes its work, will you take a walk in the area where it operated?

Exercise 1.14 An engaging book by Braitenberg[15] describes a series of fanciful vehicles that incorporate only simple sensing

---

[15]See *Vehicles, Experiments in Synthetic Psychology*, Valentino Braitenberg, MIT Press, 1984.

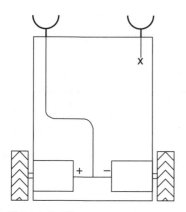 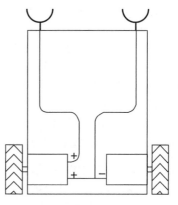

Figure 1.10

Braitenberg's type-2 vehicle uses only light sensors and motor-driven wheels. The sensors produced a voltage in response to light. Connecting the sensors to the motors in different ways produces different behaviors in response to light. The wire from the sensor can be connected positively (indicated with a "+") to make the motor turn faster with more light or negatively (indicated by "−") to make the motor turn more slowly with increasing light. (In both cases, the motors turn only forward—the negatively connected motor turns at maximum speed when the sensor detects no light.) Two (not necessarily useful) wiring schemes are shown.

and actuation. See **Figure 1.10**. How would you wire a Braitenberg vehicle to make it move directly forward any time a localized source of light is present? How would you wire the vehicle to make it move quickly in the dark and slowly in the light? How would you make the vehicle home on the light? How might the vehicle avoid the light? Can you think of a way to make the vehicle home on the light but stop some distance away? (You may have to use an additional element; remember that the output of the light sensor is just a voltage.) Can you wire or otherwise redesign a vehicle in such a way that it will orbit the light?

Exercise 1.15 Describe the motion that each vehicle in **Figure 1.10** makes in conditions of maximum and minimum light in a space that is uniformly lighted; that is, the light intensity is the same in all directions.

Exercise 1.16 Suppose that you have been given the job of implementing an entertainment robot at an amusement park. The actions of the robot mostly follow a preprogrammed script but the park manager insists that the robot only become active when a child under the age of 10 approaches the display case containing the robot. Clearly, the question you would like the robot's sensors to answer is, "Is there a child under the age of 10 near the display case?" But what questions do you actually get to ask? Can you think of a sensor, a set of sensors, or some other method using existing technology that will allow the robot to function in the way the amusement park manager wants?

Exercise 1.17 The owner of a parking garage wants to minimize costs by implementing the maximum level of automation possible. The idea is that a customer drives his or her car onto a special spot, exits the car, walks over to a ticket-dispensing machine, and presses a button. A claim check is issued and at the same time, the now vacant car is lifted by machinery into the bowels of the building where it is packed cheek-by-jowl with the other cars. The tricky part is that there is supposed to be no attendant watching the customer exit the car. Can you think of a sensor, a set of sensors, or an autonomous method that can make this operation safe? Is there a way to determine that neither driver, passenger, nor anyone else is near the car before the car-lifting machinery is activated? Will your method work for all vehicles below a certain size? What if the customer is accompanied by a small child or has a pet on a leash?

Exercise 1.18 Regarding a real-world version of the Collection task, how many ways can you think of in which the difference between photocell signals *would not* code for the direction to the light source? What could you change or add to the robot to make the robot home properly under these circumstances?

2
Control Systems

To perform a task, a robot must typically pursue various sorts of goals. An example of a high-level or task-level goal might be: "Find a land mine and dig it up." A low-level or system-level goal could be: "Maintain a forward velocity of three miles per hour." Certain goals are best pursued using a software (or hardware) construct called a control system. A control system works to maintain or achieve some condition using zero or more sensors and one or more actuators. A feedback control system uses an algorithm (or a physical mechanism) to transform input signals and the measured response of the device being controlled into new control commands.

Open and Closed Loop Control

The two broadest categories of control systems are called *open loop* and *closed loop* control systems. In an open loop system, we issue a command and hope that the system obeys it. In the example shown in **Figure 2.1**, the input to the control system is the desired robot velocity, say, three feet per second. The control system's mechanisms convert the "three" into an electrical signal that is then applied to the robot's motors. The output of the system is the robot's actual velocity.

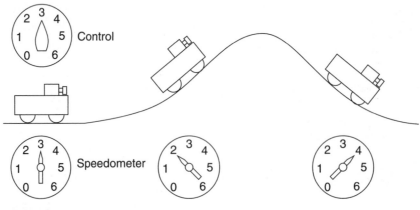

Figure 2.1

An open loop control system cannot respond to changes in the environment. Here the internal variables of the system controlling the robot have been adjusted to cause the robot to move at the desired speed when the robot rolls on level ground. But if the load on the robot's drive system increases, as it does when the robot climbs the hill, the robot slows down and so fails to move at the desired velocity. When the robot rolls down the far side of the hill, the measured velocity is higher than the desired velocity.

On flat ground, the electrical signal delivered to the motors causes the robot to move at three feet per second as desired. But when going uphill, the motors need to work harder if the robot is to maintain its velocity. An open loop system, however, doesn't consider what actually happens after it sends a signal to an actuator. If environmental conditions differ from what the open loop control system anticipates, then the robot will not behave as the designer wants.

An open loop controller is the simplest sort of control system. See **Figure 2.2** for an example. Open loop control may be sufficient if you can be sure that your robot will never encounter conditions of varying load, where it will sometimes need to work harder and sometimes ease up. Usually you can be sure of just the opposite. Making your robot perform as you want, even when conditions change, calls for a closed loop control system. A closed loop system *does* pay attention not just to what you told the robot to do, but also to what the robot actually does.

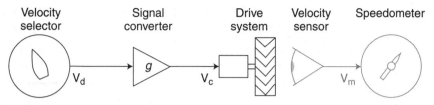

Figure 2.2

The drawing represents pictorially the open loop control system used by the robot in Figure 2.1. The velocity selector tells the robot the desired velocity, v_d, at which the robot should move. The signal converter scales the signal from the velocity selector to produce a velocity command, v_c, for the drive system. The drive system includes the motors and wheels that convert an electrical signal into the physical motion of the robot. The velocity sensor measures the velocity, v_m, at which the robot actually moves and the speedometer displays that measured velocity. (These last two components, in gray, appear here only for our later convenience; neither sensor nor display device serves any function in an open loop control system.)

As with many things that have been studied for a long time, there are a bewildering variety of types of control systems. (Perhaps the first task-achieving feedback control system was an automatic egg incubator built by Cornelis Drebbel of Holland around 1624.) We will consider one of the most common and most useful closed loop controllers, the proportional feedback controller.

Suppose that you have available a calibrated speed dial to adjust the power to the motors as in **Figure 2.1**, and a speedometer that displays the actual velocity of the robot. When you move the dial to the "one" position (and the robot is on flat ground), the robot moves at one foot per second. Twisting the dial to the "two" position makes the robot move at two feet per second and so on.

Traveling at three feet per second with the dial set at the "three" position, all is well—until you get to a hill. As the robot begins to climb the hill, you find that the speed dial and the speedometer no longer match. The speed decreases to, say, two feet per second. Anyone who has ever pedaled a bicycle up a slope will readily appreciate what has happened—to maintain a given velocity, more power is needed to go up the hill than to travel across a flat area. The *load*, how hard or easy it is for the motor to turn, has changed. The "velocity selector" in the diagram is actually a misnomer; what this control really determines is the voltage applied to the motors. The numbers on the "velocity selector" and numbers on the speedometer match only for one load. If the robot's environment changes (the load becomes higher or lower), then the robot speeds up or slows down, respectively, and the numbers no longer match.

Were you in control of the robot as it struggled up the hill, an obvious action to keep the speed constant would suggest itself. Watching the speedometer, you would ignore the "calibrated" speed and twist the dial to higher numbers until the speedometer-indicated speed matched your desired speed. Perhaps when the speed dial indicates 4.5, the robot actually moves at three feet per second as it ascends the hill. In this example, you have "closed the loop." Watching the output of the system (the measured velocity, v_m), you are able to adjust the input to the system (the velocity control signal, v_d) such that the output matches the desired velocity. A closed loop control system can do all this automatically.

In **Figure 2.3**, some additional complexity has been added to the open loop controller to accomplish closed loop control. The velocity sensor that measures the actual speed of the robot is

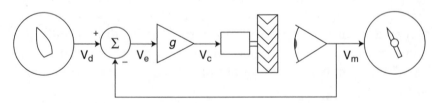

Figure 2.3

A closed loop velocity controller adjusts the velocity control command, v_c, in such a way as to reduce the difference between the desired velocity, v_d, and the measured velocity, v_m. The difference between the measured and desired value of the velocity is the error, v_e. The computation $v_e = v_d - v_m$ is performed in the circle labeled Σ. The error signal is multiplied by a gain, g, to produce the velocity command. A closed loop proportional controller can cause the robot to move at a velocity close to the desired velocity even when the load on the system changes.

now connected to something. There is a circle (labeled Σ) with two lines going in and one coming out. In the circle, the incoming signals are combined: v_m, the line marked with a "$-$" is subtracted from v_d, the line marked with a "$+$." Emerging from the circle is v_e, the error signal, $v_e = v_d - v_m$. The error signal next goes to the triangle labeled g. Here we multiply the error by gain, g, giving us $v_c = gv_e$. This value, v_c, goes to the drive system. In control system parlance, the desired velocity, v_d, is called the *reference* or *setpoint*. The drive system is called the *plant*.

With this closed loop control system operating, the robot performs correctly when it gets to the hill. As it ascends the hill, the robot's velocity begins to decrease. The velocity sensor measures the now-lower velocity and feeds it back to the *summing point*, the circle labeled Σ. The difference between the measured and desired velocity increases so that the error signal becomes larger. The increased error signal is multiplied by the gain and the command to the drive system increases. More power flows to the drive motors and the robot speeds up. When the robot descends the hill, the opposite happens. The measured velocity is higher than the desired velocity, the error signal decreases (and may even become negative), and the velocity control command decreases (or becomes negative). The motor now works to slow the robot down.

Note that unless the desired velocity is set to zero, the robot never goes at *exactly* the desired velocity. If the robot traveled at v_d then we would have $v_e = v_d - v_m = 0$. But if the error is zero, then v_c is also zero and when the velocity command is zero, the motor doesn't turn (unless some outside force acts on it). By making the gain, g, large, we can force the difference between measured and desired velocity to be as small as we like, but we can never make it zero.

You might imagine that we could fix this problem of never going exactly the right velocity by adding a constant to the velocity command to cancel the offset, but this doesn't help very much. This approach only makes the robot go at exactly the right speed for one value of load—the same load that would make the robot behave properly if it were using an open loop rather than a closed loop control system. The small offset between the desired and measured velocity is simply a property of proportional controllers. It takes a more complex controller to erase the last little bit of error.

Position Control Example

Using BSim, it is difficult to clearly observe a control loop that affects velocity, but it's easy to see what's going on in a loop that controls position. BSim includes an example of a simple control system that adjusts the position of the robot based on the intensity of the light that the robot senses. The control system is embedded in the Gizmo task. Follow the instructions linked to the BSim page to setup and adjust the Gizmo parameters.

Figure 2.4 illustrates how Gizmo works. Place the robot some distance from the light source but pointed toward the light. Then

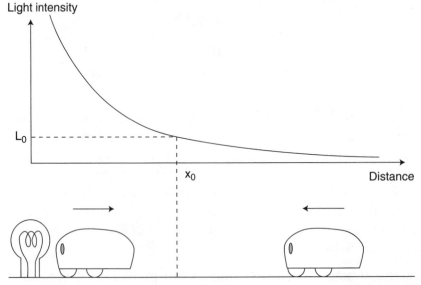

Figure 2.4

The robot uses a closed loop control system to find the point, x_0, where the light intensity, L_0, is not too bright and not too dim. If the robot starts to the left of x_0 where the light is too bright, the robot backs up. If the robot starts to the right of x_0 where the light is too dim, the robot moves forward. The robot stops moving when it reaches x_0, the point where the light intensity is L_0.

run the simulation. The robot's goal is to find the point where the measured brightness of the light is L_0. The total light, L_T, that the robot senses is the average of the light sensed by the left and right photocells, that is $L_T = (L_L + L_R)/2$. If we have $L_T > L_0$ then the light the robot sees is too bright. The robot is too close and should move backward. But if $L_T < L_0$ then the robot is in a spot where the light is too dim and so should move forward.

The difference between the desired light intensity and the actual intensity is the error, e. This is simply $e = L_0 - L_T$. When the robot is too close, e is negative; when the robot is too far away, e is positive. We can use the computed error to drive the robot to the right place if we make robot velocity, v, a function of the error. We can let $v = ge$. The full expression the control system uses to adjust the velocity of the robot is thus: $v = g (L_0 - (L_L + L_R)/2)$.

We can assume that the robot feeds the velocity command, v, into a velocity control system such as described in the previous section. When we tell the robot the velocity, we want the control system that adjusts the signal to the motors to make the robot run at the desired speed. The robot tries to get to the correct spot as quickly as possible by turning the velocity up very high when the robot is far from the right spot and reducing the velocity to 0 as the robot gets close to its destination. You may have noticed that the simulated robot in the Gizmo task closes in on the stopping point rather slowly. We have a simple way to make the robot close in more quickly—just increase the gain, g.

Control System Catastrophe

Let's try to make the control system more responsive so that we get to the desired position more quickly. Follow the instructions linked to the BSim page for increasing the gain and run the Gizmo simulation again. Does the robot behave as you expected it would? The robot certainly moves toward the goal position more quickly, but doesn't stop when it gets there. What has gone wrong?

This example exposes the dark side of closed loop control systems—sometimes a closed loop controller makes things worse.

There are many ways that this can happen, but one way that you will most commonly encounter in robotics (and the one that is built into BSim) is latency. (Open loop control systems typically are not similarly affected by latency.)

The equation our position control system uses, $v = g(L_0 - (L_L + L_R)/2)$, assumes that everything happens at the same time. We measure the light intensities at the photocells in zero time, we compute the required robot velocity in zero time, send the velocity signal to the motors in zero time, and the lower-level velocity control system adjusts the motors to that velocity in zero time. In the real world, nothing happens instantaneously.

What actually takes place is this. Let's assume that at time $t = 0$, the robot is in motion toward the light source. The robot's microprocessor then commands that a light intensity measurement be taken. The first thing the robot has to do is to digitize the analog signals from the photocells. Analog to digital converters require some number of microprocessor clock cycles to operate, so the digital value of the light measurement is not available to the microprocessor until time $t = a$ when the conversion finishes. By that time, the robot has moved forward a distance of va. The program running on the robot can't do everything at once, but rather loops through the various blocks of code. Part of the loop typically gathers the sensor values and stores them at a particular spot in the memory, time b passes while that is going on. Next to run is that part of the program that uses the sensor value to compute the velocity; time c ticks by. After that, another part of the program runs that deals with sending commands to the motors and more time, d, passes. BSim's virtual robot is finished at that point, but in a real robot, there is at least one more step. To change the robot's velocity, physical forces must be overcome. A further mechanical time delay occurs before the robot actually begins to move at the new commanded velocity, time f.

When the robot finally completes acquiring sensory information and acting on the control system's command, it will be at a point that is a distance $v(a + b + c + d + f)$ closer to the light than when sensing started. **Figure 2.5** illustrates the sequence of events. It may be the case that the robot has moved from being

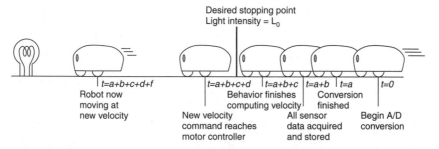

Desired stopping point
Light intensity = L_0

| $t=a+b+c+d+f$ | | $t=a+b+c+d$ | $t=a+b+c$ | $t=a+b$ | $t=a$ | $t=0$ |

Robot now
moving at
new velocity

Behavior finishes
computing velocity

New velocity
command reaches
motor controller

Conversion
finished

All sensor
data acquired
and stored

Begin A/D
conversion

Figure 2.5

Latency in the system controlling the robot causes the robot to overshoot the desired position. At $t = 0$ with the robot moving at high velocity toward the light, the robot begins to acquire the sensory information on which it will base its next velocity command. After several intermediate computations at time $t = a + b + c + d + f$, the robot finally reaches the velocity that would have been appropriate at $t = 0$. By this time, however, the robot has passed the desired stopping position. The robot's velocity command will have the wrong sign for one more iteration of the control system.

too far away to being too close to the light source. But at the moment the sensory readings were taken, the robot needed to move closer to the light—so even though it has now gone too far, the robot issues another command to move forward. Now the robot needs to back up and go in the other direction. But since the robot has only just started another round of sensing, it won't know that it should go the other way until the same time $(a + b + c + d + f)$ has passed again. All the while the robot is moving in the wrong direction. When the control system gain is high, this effect can make the robot rush from one side of the equilibrium point to the other—each cycle taking the robot further from the desired goal. This behavior is diagrammed in **Figure 2.6**.

Control System Stability

A control system that oscillates in the way we have just seen is said to be unstable. What then is the easiest way to drive the system toward stability? Every proportional controller has an associated constant of proportionality or gain. The BSim example showed us a practical way to achieve stability—you can always halt oscillations if you set the gain low enough. When the gain

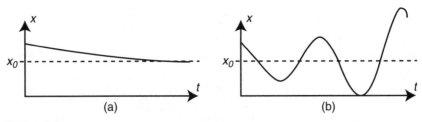

Figure 2.6

Ideally, a control system, such as the one described in Figure 2.5, converges on the position x_0 implied by light intensity L_0 as shown in (a). If the gain is set too high, however, the system may experience (possibly growing) oscillations and never settle on the correct position.

is lower, the system will not overshoot so far and any oscillations that do occur will tend to damp out. The only mischief that reducing the gain brings on is that the system does not reach the goal point as quickly as would a system with higher gain (also, a larger offset between desired setpoint and actual output may remain). This presents no problem in our simple light-seeking example; there is no particular urgency in getting to the equilibrium point. But you can likely think of many examples where a sluggish control system would cause trouble.

At the cost of increasing the complexity of the system (and complicating the mathematical analysis), there are ways to zero in more quickly on the goal without generating diverging oscillations. One general type of closed loop controller is called a PID controller—a generic example is shown in **Figure 2.7**. PID stands for proportional, integral, differential. We have already seen the proportional part. In the proportional branch, the control signal is proportional to the error. A differential controller includes a term that is proportional to the derivative of the error signal; that is, how rapidly the error signal changes with time. An integral controller adds a term that is proportional to the integral of the error signal, the summation of the error signal over time. (The integral branch allows us to fix the proportional controller's problem of never achieving exactly the commanded value.)

The complexity of a PID controller is not just three times that of a P controller. Like the P branch, the I and D branches also have an

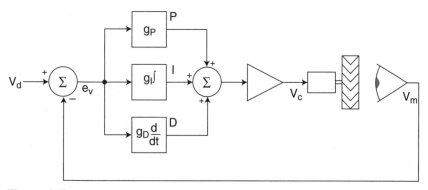

Figure 2.7

A general PID control loop includes a differential and an integral term as well as a proportional term. The D or differential branch looks at how quickly the error term is changing and modifies the error signal based on the rate of change. The I or integral branch integrates (adds up over time) the small offset uncorrected by the other branches and uses this to cancel out offset errors. Each branch has an associated gain: g_P, g_I, and g_D.

associated gain. To achieve good performance, each gain must be adjusted—we call this *tuning* the system. The idea is that the control system should respond rapidly to disturbances (changes in the load), cancel out any leftover constant offsets, and not oscillate wildly in the process. It is easy to tune a P controller—adjust the gain until the system converges on the desired setpoint as rapidly as possible without generating oscillations. There is only one knob to turn. Not so with a PID controller. This controller has three knobs to turn, but here's the problem: *you can't tune the system by optimizing each branch independently!* Cross-coupling within the system means that adjusting the gain of one branch affects the optimal settings of the other branches. And, despite years of mathematical analysis, getting the best real-world performance from a PID controller still remains something of a black art.[1]

---

[1]A thorough understanding of PID controllers requires familiarity with differential equations. Studying PID controllers, you will learn about imaginary numbers, pole-zero maps, Laplace transforms, and other nefarious constructs you'd do well to avoid running into in a dark alley. A popular and complete, if not especially novice-friendly, book on control systems is: *Schaum's Outline of Feedback and Control Systems* by A. Stubberud, I. Williams, J. DiStefano, McGraw-Hill, 1994. Another perhaps more easily understood book is *Feedback Control of Dynamic Systems*, 3rd ed., G. Franklin, J. Powell, A. Emami-Naeini, Addison Wesley, 1994.

It doesn't stop there. Control systems are an active area of research and development. Over the years, researchers have devised all sorts of exotic controllers: feed forward controllers, predictive controllers, adaptive controllers—more types of controllers than you can shake a stick at. Thankfully, these Rube Goldbergian structures are beyond the scope of this book. We sacrifice little by leaving out the more exotic controllers. At the lowest levels of the hardware, you can use all the control system sophistication you can get. But at the higher levels of behavior-based programming, simple, proportional controllers nearly always suffice.

Even those who develop a thorough understanding of control systems and the differential equations that govern them know of no magic that forever banishes control system catastrophes. Each time you build and program a robot, you will confront anew the limitations of sensors, actuators, and control systems. A deeper understanding will, however, allow you to build more sophisticated systems and achieve acceptable performance more quickly.

Saturation, Backlash, and Dead Zones

Even allowing the problem of latency and stability, there's still something a little too pat about our light-to-position controller example. In a couple of ways the computed response of our controller doesn't quite tally with physical reality.

It is always a good practice in robotics (as it is in other fields) to consider the practical consequences of any equation you have just solved or any algorithm you are about to implement. In our proportional controller, we decided that the velocity command, v, should be $v = g(L_0 - (L_L + L_R)/2)$. What happens at the extremes? When the gain is high and the robot is far from the correct position, the computed value of v can become quite large. In fact, the commanded velocity can easily exceed the highest speed at which the robot is able to travel. What happens when the controller tells the robot to do something that the robot is physically unable to do?

You guessed it—the robot disobeys the controller. At some point the system reaches *saturation*. You can command the robot to go faster than the saturation speed, but the robot will refuse and instead stubbornly plod along at its maximum velocity. Sometimes this mismatch can cause problems.

There is another rather subtle problem that a controller may exhibit at extreme low velocity. As the robot gets closer and closer to the correct position, the velocity command gets smaller and smaller. At some small but not yet zero-commanded velocity, the motor stops turning. If you are using a high-quality motor and a simple proportional controller, this event is not a major concern. (The virtual motors BSim uses are of the highest quality—they continue to spin slowly even when infinitesimally small voltages are applied.) But if your physical motor is a more typical (and more affordable) one, internal friction will make the motor stop turning when the robot is some distance from the desired setpoint.

Zero measured velocity in the presence of non-zero commanded velocity means that there is effectively a dead band or dead zone around zero commanded velocity—below some minimum, commanded velocity the system will not respond. **Figure 2.8** compares the ideal system response with a more realistic model including saturation and a dead zone.

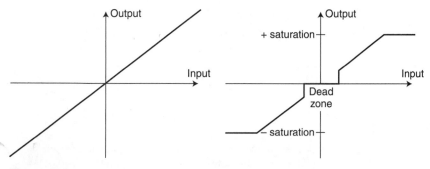

Figure 2.8

We might imagine that systems respond linearly to control commands as shown by the graph on the left, but the graph on the right is closer to reality. Every system has limits. Beyond the system-specific saturation limit, system output is constant regardless of the input command. Many controllable systems, motors for example, also have an unresponsive region, or dead band, near zero.

Often (especially when your controller is of the more sophisticated type that includes an integral term) it is helpful explicitly to build saturation and a dead zone into your controller. By so doing, you avoid asking the robot to perform actions that it is unable to perform; this can eliminate one source of instability. (Pseudocode implementing a simple algorithm that accounts for saturation and a dead zone is presented as an example in Appendix D.)

Actuator systems exhibit at least one more form of rebelliousness toward their controllers; the treachery is caused by backlash. In robots, motors are almost always accompanied by gear trains. An example is shown in **Figure 2.9**. Gear trains necessarily have small gaps between the teeth of adjacent gears. Thus, when the motor drives several gears in series, the last gear in the sequence does not begin to turn at the same instant the motor starts turning. Before the final gear can turn, all the small gaps between the teeth of all the gears in the sequence must close up. When the motor switches direction, the output gear hesitates

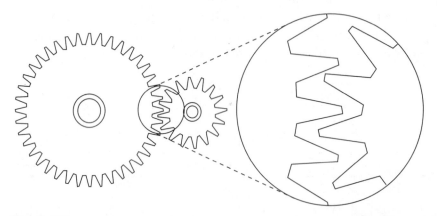

Figure 2.9

Gears are mechanical devices manufactured to certain tolerances. Even the highest-quality gears need small gaps between the meshing teeth to avoid binding and excessive friction. A magnified view of the meshing region is on the right. Suppose that the small driving gear begins to spin clockwise. The larger, driven gear does not begin to spin until the gap between the small and large gear teeth has closed. If the driving gear then reverses direction, the driven gear stops spinning, the gap between the teeth closes in the other direction, and then the driven gear begins to spin the opposite way.

again as all the gaps close in the other direction. This gap-closing effect is known as backlash, and it can sometimes cause trouble for the system controlling the actuator. (You may have observed an example of backlash if you have ever tried to set a mechanical clock. Reversing the direction of the setting knob does not immediately reverse the direction of motion of the clock hands.)

Consider what happens when a control system attempts to move a backlash-plagued actuator to a particular position. The control system notices that the actuator is in the wrong position and a command is given to the motor to turn. The actuator does not immediately begin to move. The control system's error term thus increases, causing more current to flow to the motors. Suddenly the last gap closes and the final gear begins to spin. But now the motor is spinning too fast, so the actuator overshoots the desired position. The control system then commands the motor to reverse direction and move the actuator back. And again, the gaps open, then close and once again the actuator overshoots. The result is chatter—the control system constantly moves the actuator back and forth to opposite sides of the desired position. Adding a dead band to the input side of the controller that corresponds to a distance (or angle) larger than the sum of the gaps is one way to mitigate this problem. This enables the controller to ignore small errors.

Open Loop Controllers with Parameters and State

To qualify as open loop, all a control system need do is ignore the effect its commands have on whatever is being controlled. The control signal produced by an open loop controller can otherwise be dynamic and arbitrarily complex. Let's consider an open loop controller that changes its output over time to accomplish a task.

Suppose we have a robotic train that runs on a closed track through a field of berry bushes. The robot's purpose is to discourage birds from eating the berries. In order to scare the birds away, we want the robot to turn on every once in a while and drive along the track making as much noise as possible. The robot shouldn't move constantly because the birds might become habituated to the motion and ignore the robot.

To have the system properly frighten the birds, we'll establish two time intervals, t_{on} and t_{off}, that correspond, respectively, to the number of seconds the robot runs along the track and the number of seconds it sits idle. When the controller is first activated, it sets the velocity, v, of the drive system to some constant value, v_0. After the time t_{on} has passed, the controller sets the velocity to zero, $v = 0$. When an additional time t_{off} has gone by, the controller again sets the velocity to v_0 and the process repeats. The robot thus moves in a way that is controlled by three numbers: v_0, t_{on}, and t_{off}. These numbers are *parameters* of the control system. By adjusting the three numbers, we can make dramatic changes in the way the robot operates. As we saw before, proportional control systems always have at least one parameter, gain. The adjustment of the gain has a strong effect on the way that proportional controllers function.

Later we will see that it is useful to analyze controllers in terms of states. The bird train has two states, On and Off; this is diagrammed in **Figure 2.10**. It also has rules for switching between the states—when t_{on} seconds have passed since the robot entered the On state, it switches to the Off state. After t_{off} seconds in the Off state, the controller switches to the On state.

Bang-Bang Controllers

Another uncomplicated type of controller that sees frequent service in mobile robotics is the *bang-bang controller*. Eschewing the subtle sophistication of proportional control, the bang-bang controller has just two states. But unlike the bird train example, switching between the states depends on feedback. A bang-bang controller typically outputs one value if the measured quantity of interest is above a selected threshold and a different value if that quantity is below the threshold.

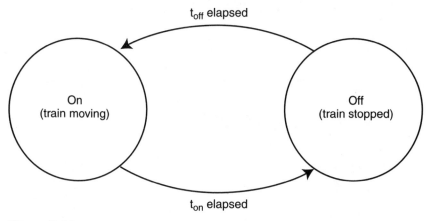

t_{off} elapsed

On
(train moving)

Off
(train stopped)

t_{on} elapsed

Figure 2.10

The bird-frightening train has states On (where the train is moving) and Off
(where the train is stopped). After time t_{off} has elapsed in the Off state the
system switches to the On state. When time t_{on} has passed in the On state the
system changes to the Off state.

Although simple, needing only one bit of input, a bang-bang
controller can be quite effective. We can use such a controller to
enable our robot to follow a wall. See **Figure 2.11**. In this example, the robot tries to follow a wall that the robot expects to find
to its right. The robot senses the wall using an IR proximity
detector. The output of such a proximity sensor is just a one or a
zero. One means that the wall has been detected; zero means that
there is nothing there. To follow the wall, all the robot has to do
is describe a circular arc with origin on the left or right of the
wall depending on whether the sensor does or does not detect
the wall.

System latency (how quickly the system senses and reacts)
determines in part how well the robot follows the wall. A large
amount of latency will generate a path that consists of large scal-
lops. A system with little latency allows the robot to move in a
nearly straight line making barely perceptible corrections.

Although this bang-bang controller has only two states, it still
possesses a gain parameter, the radius R, of the arc the robot follows. For large values of R, the robot can follow the wall smoothly even when the system exhibits a large amount of latency. But

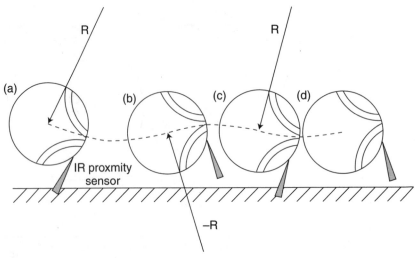

Figure 2.11

This wall follower knows only two states—too near and too far. When the robot senses the wall, (a), the robot arcs away from the wall describing a circular arc of radius R whose origin is located left of the robot. After the robot has moved far enough away from the wall, (b), that the sensor can no longer detect the wall, the robot arcs back toward the wall. Now the robot follows a circle of radius R whose origin is located to the right of the robot. The robot continues in this manner alternately arcing toward the wall, (c), and away from the wall, (d). (In robotics, if not in mathematics, it may be helpful to think of the robot as moving about a circle of negative radius, $-R$. See the discussion for Figure A.7 in Appendix A.)

if the wall makes a sharp turn, the robot may bump the wall or move a good distance away while attempting to return to the wall. With small values of R, the robot reacts quickly to variations in the wall, but may spin wildly back and forth if system latency is large.

Hysteresis

An important concept suggested by the bang-bang controller is the notion of *hysteresis*. Consider a thermostat used to control the temperature of a room. When the air in the room is too cold, the thermostat turns the furnace on. When the room becomes too warm, the furnace goes off. Such a control system might be implemented in pseudocode in this way:

```
Function furnace (T, T₀)
    If (T < T₀)
        Turn_furnace_on
    else
        Turn_furnace_off
    end if
end furnace
```

In this simple approach, the thermostat is balanced on a knife edge. A micro-degree too cold and the furnace activates, a micro-degree too hot and it shuts off. The furnace thus cycles continuously. For some devices (inexpensive battery-powered motors for example), rapid on-off cycling is acceptable—it is even encouraged as a clever way to modulate speed. But for other devices like a furnace, rapid cycling leads to inefficient operation and reduced life.

Figure 2.12 presents an example solution. Hysteresis can be deliberately introduced into the furnace's bang-bang control system such that the furnace turns on at one temperature but does

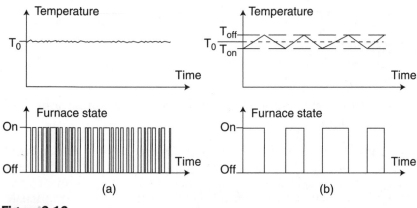

(a)   (b)

Figure 2.12

The behavior of a thermostat-controlled furnace without deliberately introduced hysteresis is plotted in (a). The desired temperature is T_0. Tiny variations around this value cause the furnace to cycle on and off at short intervals, possibly damaging the mechanism. In (b), hysteresis has been deliberately introduced. The furnace turns on when the temperature falls to T_{on} but does not turn off until the temperature climbs to T_{off}. Cycling is reduced at the cost of allowing the actual temperature to vary farther from the desired temperature.

not turn off until a higher temperature is reached. Moderated by the slow change of the temperature of the room, the system thus cycles less rapidly. Such a system might be implemented in this way:

```
Function furnace_hys (T, T_0, ΔT)
    If (T < T_0 − ΔT)
       Turn_furnace_on
    else if (T > T_0 + ΔT)
       Turn_furnace_off
    end if
end furnace_hys
```

When the temperature is in the dead zone between T_{on} ($T_{on} = T_0 - \Delta T$) and T_{off} ($T_{off} = T_0 + \Delta T$), the furnace remains in its previously commanded state, either on or off.

A small amount of hysteresis is inevitable in any system (latency can be a source of hysteresis). Depending on the purpose of your control system, optimal performance may demand larger or smaller values of hysteresis.

Summary

In this chapter we have learned that:

- Control systems come in two basic flavors, open loop and closed loop.

- Open loop controllers are simple and stable but cannot respond to environmental changes.

- Closed loop controllers adjust the robot's performance to deal with outside disturbances or changing conditions.

- Closed loop controllers can be unstable (oscillate) unless properly adjusted.

- Simple bang-bang controllers have just two states but can often be effective.

- Control systems have one or more parameters that affect their performance.

- A robot is not always able to execute the commands computed by a control system.

Exercises

Exercise 2.1   Consider the Gizmo task. Is there a way to write an open loop control system that will move the robot to the correct position relative to the light source?

Exercise 2.2   Think of an example where rapid convergence of the control system (getting to the setpoint without oscillation) is a requirement. What happens if the requirement is not met?

Exercise 2.3   In the bird-frightening example, how would you adjust the parameters to scare away birds with a short attention span? How about birds with a long attention span?

Exercise 2.4   It was stated above that a proportional feedback control system makes the robot go at *almost* the commanded velocity. Why is it almost rather than exactly the right speed? Express mathematically the velocity error that remains uncorrected even when the system is acting properly. (Hint, find an expression for v_m and compare it to v_d.)

Exercise 2.5   Maybe it looks as though it would be a good idea to add a constant term to the proportional controller. We could arrange the system a bit differently such that $v_c = g_1 v_d - g_2 v_m$. With this setup we would move at exactly the desired velocity whenever the error term is zero. Think carefully about the situation and explain why, in general, this extra complexity has little benefit.

Exercise 2.6   We didn't say much about how the control system plant (the robot's drive system) responds in our velocity control example. But clearly, the output velocity depends on the control signal and the load. Let's model the plant's behavior by writing: $v_c = a\, v_c\, (L_o/L)$, where L is the load and L_0 is the initial load. If the plant responds in this way, what will v_c be when the robot is going up hill if $L = 2L_o$? What will be the error velocity in that case? If $a = 1$, and latency is 1 second, predict the value of g that will make the system go unstable. (Very hard).

Exercise 2.7 Besides the examples we have seen in this chapter, what other sorts of robot variables might we want to manage with a control system?

Exercise 2.8 Is a flush toilet (one with a toilet tank) an example of an open loop or closed loop control system? What quantity does such a system control?

Exercise 2.9 A stepper motor is an electric motor that advances one step (step size is a specification of each motor model) when one electric pulse is sent to the motor. Stepper motors are frequently used in robotics to move some device to a particular position. Assume that we are using a stepper motor to position a miniature video camera. To ensure that sending a given number of pulses to the motor will bring the camera to a known position, the motor is typically driven into a hard stop at power up. That is, when the power turns on, the system sends enough pulses to be sure that the motor has reached its travel limit. The camera is then assumed to be at the zero point. Should such a positioning system be classified as open loop or closed loop? What sensing mechanism(s) would you need in order to properly position the camera if you used a more common type of motor, one that is not a stepper motor? Would control of such a system require a closed loop controller or could an open loop system be made to work?

Exercise 2.10 Imagine that the Gizmo task is arranged a bit differently and that the light is mounted on top of the robot rather than being fixed in the environment. Suppose further that the robot's photocells are shielded in such a way that the photocells cannot see the robot-mounted light directly but can only see the light reflected from a wall or other object. Describe the behavior of the robot in this situation. If we keep the same gain setting as we had when the control system was optimized for the fixed light, is the system now likely to be more stable or less stable? Why? How is the setpoint (the distance from the wall or object to the point at which the robot stops moving) affected by the color of the surface? How is it affected by the size of the object?

Exercise 2.11   Describe a system for controlling the altitude of a hot air balloon. Can you use a bang-bang controller for this application? Identify the quantity (or quantities) that correspond to the gain in this example.

Exercise 2.12   Besides the home heating system mentioned in the text, what other control systems might benefit from hysteresis? Which control systems would suffer?

3

Behaviors

In the previous chapter, we learned about basic control systems. A control system is a key element in any robot program, but a simple control system turns the robot into a sort of one-trick pony—making the robot act in the same way all the time. To perform useful work in the real world, we must have our robots do different things under different circumstances. And here enters the concept of behaviors.

Triggers and Control Systems

Recall the example from the first chapter where we used the simulated robot to carry out the Collection task. One of the operations that the robot performed was homing on the light source. To home on the light source the robot ran a light-seeking control system. But the robot did not spend all of its time homing on the light source. Rather, the robot performs the homing operation only when it is pushing a puck. The presence of the puck *triggers* the homing behavior.

Primitive behaviors, as we use the term in behavior-based robotics, have two parts:

1. A control component that transforms sensory information into actuator commands.

2. A trigger component that determines when it is appropriate for the control component to act.

Figure 3.1 diagrams a generic behavior.

In our Collection example, the absence of a puck means that the robot should not home on the light—when the robot is not pushing a puck, it should go out and find one to push. Without a puck, the primitive homing behavior is untriggered. But when a puck is in contact with the robot's bumper, the homing behavior is triggered and the robot tries to move toward the light source. In this case, the trigger part of the behavior looks at the bumper sensor to decide if homing is appropriate, and the control system part of the behavior looks at the photocells to determine how to home.

As far as the system outside the primitive behavior is concerned, it doesn't matter whether the absence of a trigger signal

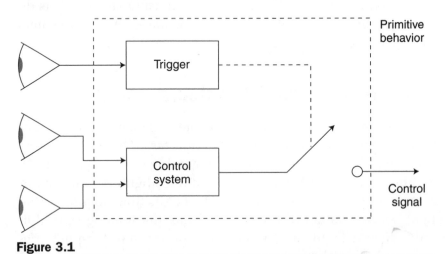

Figure 3.1

A primitive behavior includes a control component (a control system) that maps sensory information to actuation control commands and a trigger to determine the circumstances under which the control system should generate commands for the robot's actuators. Both control system and trigger can examine sensor values to decide what to do. The trigger and the control system may consult the same sensors or different sensors.

stops the control system from computing commands or only prevents the control system from sending commands to the actuators. (The latter case is indicated by the dashed line in the figure). The system as a whole behaves in exactly the same way regardless. Sometimes, allowing the control system to run constantly rather than turning the computation on and off can achieve some utility or program simplification. If the control system runs constantly, the trigger-controlled switch simply discards the output commands when they are not appropriate to the situation.

Servo and Ballistic Behaviors

A primitive behavior is a general construct that places few restrictions on the implementing code. But there are two categories that broadly describe behaviors: servo and ballistic. You have already seen examples of both types in the Collection task.

Typically, a servo behavior employs a feedback control loop as its control component. The light positioning behavior from Chapter 2 is an example of a servo behavior. Suppose that we increased the intensity of the light source when the robot is at rest at the equilibrium point. The robot would immediately respond by backing up. If we decrease the light intensity, the robot will move closer without reluctance. The trigger is always active, so that at every instant, the robot measures the light on the photocells, compares the sum with a stored intensity, and then moves forward or back depending on the difference.

A ballistic behavior, like a shell fired from a cannon, once triggered, follows a predictable trajectory through to completion. The Escape behavior is an example of a ballistic behavior. Escape executes when the robot bumps into a wall or other obstacle. When triggered, Escape does these things: First, the robot backs up a preset distance. Second, the robot spins in place a preset number of degrees. Third, the robot moves forward a preset distance. Then Escape becomes untriggered. (Step Three exists to make sure that the robot is in a different spot when another behavior assumes control. Without the move forward, whatever behavior had run the robot into the wall might just run it into the wall again.)

Usually, we do not think of behaviors as "completing." A behavior just runs continuously, always trying to achieve some goal or maintain some value. But ballistic behaviors provide an exception to this model; as in the case of Escape, they often have a clear end point. You can view a ballistic behavior as a small plan or sequenced operation.

Ballistic behaviors, although sometimes essential, must be used with caution. Depending on how it is written, during the time that a ballistic behavior runs, the robot can be effectively blind. If conditions change during behavior execution or if the robot has made a mistake in initiating the behavior (it may have been started by a noise glitch, for example), trouble can result. Before resorting to a ballistic behavior, it is always best to try first to find a servo behavior that will accomplish the same result. Servo behaviors respond immediately to changing conditions and are less vulnerable to noise and other glitches.

Implementing Servo Behaviors

What does a servo behavior look like? Servo behaviors need not be complicated. A behavior that controls the drive motors of a robot has only to compute a velocity for the motors. The details of implementing a behavior depend on the software system that runs the robot, but before writing any code, it is critical to have a clear picture of the method or algorithm you plan to use. Let's look at a couple of examples.

First, consider one of the simplest of all possible behaviors—call it Cruise. Before writing our first line of code, we need a clear description of what Cruise does. For our purposes, Cruise is a behavior that drives the robot forward at constant speed no matter what obstacles or hazards the robot may encounter.

Regardless of the robot's situation, Cruise continually does the same thing. This means that the trigger portion of the behavior has no need to monitor any sensors. In fact, there is no need for an explicit trigger at all because the trigger would always be active. Because Cruise commands the robot to go in only one direction, there is nothing for the control system part of Cruise to compute— all Cruise has to do is to output the desired velocities for the drive motors. Let's suppose that we are controlling a BSim-like robot (see **Figure 1.2**) and that the software system knows about two variables, v_left and v_right, which specify the velocity of the left and right drive wheels, respectively. The robot's lower-level control code causes the left and right wheels to move at whatever velocities are stored in variables v_left and v_right.

If all we want Cruise to do is to make the robot move forward at velocity, v, then the code that implements Cruise should set both v_left and v_right equal to v. In pseudocode, we might write this:

Behavior **Cruise**
    v_left =v
    v_right=v
end Cruise[1]

Cruise behaves much like a subroutine. We imagine that the robot's software system calls Cruise over and over as rapidly as it can (possibly several hundred times per second). Cruise ignores all sensors, always telling the robot to do the same thing no matter what happens to the robot.

We think of the variable v as a parameter of Cruise. The value of v is specified externally to Cruise and the particular choice causes Cruise to control the robot in different ways—making the robot go faster or slower. Parameters can have a dramatic effect on how behaviors direct the robot. Here is a slightly more interesting example:

---

[1]I'm cheating a bit here by ignoring the problem of exactly what v_left and v_right are. If these are global variables, then behaviors other than Cruise can alter their values. If they are variables local to Cruise, then the motor driver code can't see them. Thus v_left and v_right have to be special in a way that will be described later. For now, don't worry about other behaviors overwriting v_left and v_right and assume that the motor driver has a way to know their values.

Behavior **Cruise-plus**
    v_left =$v - b$
    v_right=$v + b$
end Cruise-plus

Cruise-plus has two parameters, v and b. When b is zero, Cruise-plus behaves in a way identical to Cruise—it commands the robot to move straight forward at velocity v. But by choosing a non-zero value for b, we can cause the robot to spin or arc as we please. If b is a positive number (small compared to v, which we assume also to be positive), then Cruise-plus will set the velocity of the left wheel to a value a little less than the velocity of the right wheel. The robot will make a shallow arc to the left. Given enough space, the robot will describe a large circle. If we set v equal to zero, and b not equal to zero, Cruise-plus will give the left and right wheels opposite but equal velocities—the robot will spin about the point midway between the two drive wheels. Thus by incorporating a couple of parameters, we have been able to add greatly to the versatility of a very simple behavior. (For a BSim experiment, see Exercise 3.8.)

Next let's try a servo behavior that may score a point or two higher on the IQ scale than Cruise or Cruise-Plus. Please visit the BSim site and activate the User Control task containing the Remote behavior. Remote provides an interface between the robot and the user, enabling the user to drive the robot directly. Remote does this by monitoring the buttons on BSim's joystick pad: F, L, R, and B. These letters represent forward motion, a spin to the left, a spin to the right, and backward motion, respectively. We will imagine that the robot's software system includes a function called pressed(x). If the user has pressed the button labeled x, then pressed(x) returns True, otherwise it returns False. (Those familiar with the C programming language may wish to think of false as being zero, and true as being any non-zero number.)

Behavior **Remote**
    If (pressed(F)) then
        v_left = v
        v_right = v

```
    else if (pressed (L)) then
        v_left = −v
        v_right = v
    else if (pressed(R)) then
        v_left  = v
        v_right = −v
    else if (pressed(B)) then
        v_left = −v
        v_right = −v
    end if
end Remote
```

Like Cruise, Remote is called, perhaps every few milliseconds, by the robot's software system. But unlike Cruise, Remote does not necessarily output any command; that is, it does not always pick a velocity for the drive motors. If no buttons are pressed, then the Remote behavior has no opinion of what the robot should do and Remote, therefore, outputs nothing. But as soon as any of the buttons are pressed, Remote's self-confidence surges, it knows exactly what the robot must do and Remote resolutely reports its choice by assigning values to v_left and v_right repeatedly, as often as the system software gives Remote a chance to run. Thus, even though Remote has no single, explicit part labeled trigger, pressing any of the joystick buttons does trigger Remote. The code that implements the trigger function is distributed throughout the Remote behavior.

It is easy to see what Remote does. When the forward key, F, is pressed, the left and right wheel velocities are both set to v, and the robot moves forward. When the L key is pressed, the left wheel velocity goes to $-v$, the right wheel to $+v$, and the robot spins in place to the left. In a similar way, the R key makes the robot spin to the right and the B key makes the robot back up. What does the robot do when no buttons are pressed—does the robot move or stop? We cannot predict by looking only at the Remote behavior. What the robot does when Remote sends no commands depends on other parts of the system. (We could, of course, add one more clause to Remote to choose to stop when no buttons are pressed.)

Now we've seen behaviors that pay attention to nothing (Cruise and Cruise-plus) or pay attention only to the user (Remote). Next let's examine a behavior that makes use of a trigger and computes a motion command based on sensor values. Examine Home, a more complex behavior, but one that conforms to the same general format. Home causes the robot home in on a light—provided that the light is bright enough.

Behavior **Home**
```
    if ((photo_left + photo_right)/2 > light_min) then
        v_left   = v + home_gain * (photo_right – photo_left)
        v_right = v – home_gain * (photo_right – photo_left)
    end if
end Home
```

To analyze Home, we must suppose that the robot's software system provides two variables, photo_left and photo_right, which correspond to the amount of light measured by the left and right photocells, respectively (the more light, the larger the value). In addition to the parameter v, the nominal velocity, the system includes two other parameters, one called light_min and home_gain. Light_min is the minimum amount of the light to which the robot will respond. Home_gain is a number that determines how strongly the robot turns in response to the difference in the light levels falling on the photocells.

Whenever the average amount of light, (photo_left + photo_right)/2, is greater than light_min, Home computes an output. If the amount of light falling on the photocells is less than this number, Home does nothing. Thus, Home triggers on light level—the behavior ignores dim sources of light in its environment and homes in only on sources that appear bright.

As you can see from the equations for v_left and v_right, if the amount of light that the left and right photocells observe is the same, then photo_right – photo_left = 0 and the robot moves straight forward at velocity v. The left and right photocells (which point diagonally forward) will see identical light levels only if the light source is dead ahead. But if the light source is a bit off to the left of the robot, then more light will fall on the left photocell than on the right. Photo_left will be greater than photo_right so that the difference (photo_right − photo_left) multiplied by the positive number home_gain will be a negative quantity. The velocity of the left drive wheel is computed by adding this quantity to *v*; to get the velocity of the right drive wheel, we subtract the negative quantity. Thus, the robot's left wheel turns more slowly than the right wheel. When this happens, the robot turns toward the left, thus pointing more directly toward the light—the robot homes in on the light.

As we saw in the control system example from Chapter 2, we can determine how strongly the robot reacts to left/right differences in light level by adjusting the parameter follow_gain. When follow_gain is zero, the robot moves when the light is bright enough but does not follow the light. When follow_gain is positive, the robot does follow the light (unless the number is too large and the system collapses into oscillations). But note what happens if we make follow_gain a negative number. In that case, rather than turning toward a bright light, the robot tends to turn away. By changing the sign of the gain parameter, we can make the robot seek darkness rather than light!

Given that the robot has only two light sensors, have we now exhausted the possible ways in which our robot can respond to light? No, we are not even close. Later on, those with the strength of heart to take on an elementary matrix multiplication will learn of even more subtle control possibilities enabled by the *general linear transform* (see the section, "Generalized Differential Response" in Chapter 5).

Finite State Analysis

Pure servo behaviors (that, like the ones above, contain no integral terms) live in the moment. Every time such a behavior is called, it computes what to do right now. The behavior pays no attention to what it did the last time it was called and it makes no preparations for what it will do next time. We say that such behaviors have no *state*.

State is a very general term that refers to the possible configurations of a system, but in the robotics context, saying that a system (a robot program for example) has state generally means that the system has memory. When a system possesses state, the system's responses are not influenced exclusively by the current values read from the sensors; actions are also influenced by sensory inputs collected and processed in the past.

Some systems have an infinite number of states. One example of an infinite state space is the arrangement of furniture in a room—you can always generate a new and different configuration by moving a chair a nanometer in one direction or another. But systems with only a finite number of states find frequent use in robotics. The *finite state machine* (FSM), is a type of system that has a limited number of states and has well defined rules specifying how the system is allowed to transition from one state to another.

We can build a very simple finite state machine, illustrated in **Figure 3.2**, that is composed of a light bulb connected to a power supply via a pushbutton switch. We will assume the switch is of the latching type—pressing and releasing the button completes the circuit, making the light go on and stay on; pressing and releasing the button again makes the light go off and stay off.

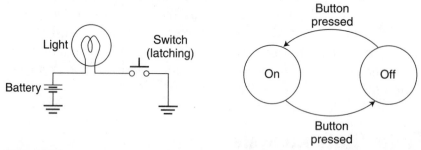

Figure 3.2

On the left is the schematic diagram of a simple electric circuit. The finite state machine diagrammed on the right describes the circuit's operation. The system, composed of a pushbutton switch, light bulb, and power source, has two states, on and off. Pressing the button constitutes an input to the finite state machine; an input can cause a transition from one state to another. The system can occupy only one state at a time. State transitions may be accompanied by an output from the system, but none is needed in this example.

This finite state machine has two states, On and Off. It has a simple rule that tells us how to get from one state to the next: press and release the button. The system also exhibits memory. The light stays on even when no one is holding down the button, so if we come upon the system and find the light glowing, we know one bit of the system's history—at some point someone must have pressed the button.

You may not yet see the utility of invoking the finite state machine in the analysis and construction of robot programs. But without this simplifying concept, even moderately complex robot programs can quickly become a mishmash of special cases and exceptions. Let's consider a slightly more complicated example—the automatic garage door opener.

Installed in my garage is a garage door opening system that behaves in the following way. If the door is closed and I press the open/close button, the door begins moving up. When it is all the way open, the door trips a limit switch and motion stops. If I push the button when the door is open, it begins moving down. As the door reaches the bottom of its travel, it trips another limit switch such that motion stops just when the door is fully closed.

There are many ways to describe the behavior of the garage door in terms of finite state machines. One fairly obvious approach is to declare that the system has four states: door closed, door opening, door open, and door closing. The system transitions between states when I press the button or when the door trips a limit switch. This is diagrammed in **Figure 3.3**. The system can remain in the open or closed state indefinitely but can occupy the opening or closing states for only a few seconds, until the moving door reaches the limit of its travel and trips a switch.

So far so good, but the actual system is a bit more complicated than described. I don't have to wait until the door reaches its fully up or down position before I press the button again. If I push the open/close button while the door is moving, the door halts immediately. Pressing the button again from this state makes the door move in the direction opposite to the way it had been going. We might choose to handle the added complexity by

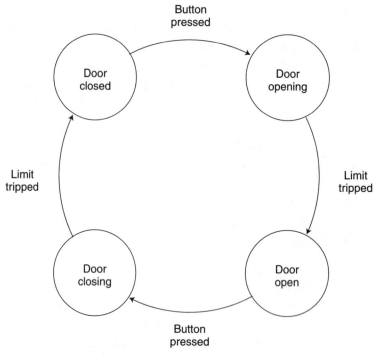

Figure 3.3

A finite state machine diagram can illustrate the operation of a garage door opening system. Here the system is decomposed into four states. When the user presses the open/close button or when the door reaches the limit of travel, the FSM transitions between states.

incorporating a few more states into our finite state diagram, as shown in **Figure 3.4**.

The complexity of our system description is building, but we are not yet overwhelmed. Although our diagram is becoming dense, it at least remains pleasingly symmetric. The original four states we defined for the system were obvious at a glance—you could tell which state the system was in by looking at the physical door. But the two new states, Stopped opening and Stopped closing, are visually indistinguishable. This may be a little disquieting, but we have arrived here logically; maybe this is the best we can do. After all, which way the door was moving before it stopped is a fact that the system must remember so that it can make the right choice when the button is next pressed.

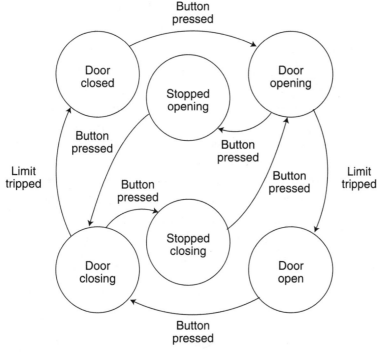

Figure 3.4

One way to deal with increased complexity in the system is to add more states to encode the complexity. Here we have added two new states, Stopped opening and Stopped closing, to more accurately reflect the system's behavior.

It turns out that there is yet a bit more complexity in the system. The garage door opener includes a courtesy light that comes on automatically as soon as I press the open/close button. The manufacturer's intent is to give the user time to get into or out of the car at night when, without the courtesy light, the garage would be dark. After pressing the open/close button, the light goes on and remains on until 4.5 minutes have passed—then a relay clicks and the light blinks off.

How shall we encode this feature into the finite state diagram of the system? (For the last subtle bit of complexity left in the garage door, see Exercise 3.12 and Exercise 3.13.) Following our example so far, you might expect that we will now have to double the number of states to 12. This is because each of our existing states needs now to include information about the light—is it on or off? We will also need to add a new state transition rule that encodes the timeout feature: the light turns off after it has been on for 4.5 minutes.

It is time to pause and reconsider. Finite state-machine analysis is supposed to aid our understanding, making it easier to write the code that implements the system. But instead, our diagram (and our thinking) is becoming unwieldy—a condition almost certain to lead to mistakes. We've arrived at this awkward point because we have tried to encode every fact we know about the system into a particular state. We are allowed, however, to store information in a variable rather than an explicit state, if that simplifies the system. Also, we are about to mix two systems, door control and courtesy light control, which are in fact mostly independent. Let's try again. We will separate the two parts of the system and add a variable, *d*, that tells which way the door should move. This is shown in **Figure 3.5**.

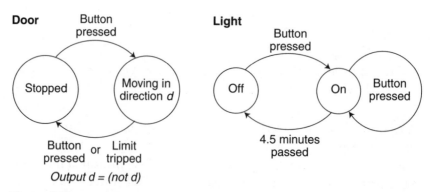

Figure 3.5

This finite state implementation of the garage door/light example offers great simplification. Operation of the door and light are separated into different FSMs with just two states each. The door is either moving or stopped, the light is either on or off. The input of a button-press or a limit switch-trip from the "Moving in direction d" state causes an output from the door FSM. The "Moving in direction d" state outputs a value for *d* that is opposite to the current value. The next time that the "Moving in direction d" state is entered, the door will move in a direction opposite to its previous direction—if it was moving up it will go down, if down it will go up, if stopped midway, the door will move in a direction opposite to the way it was going when stopped. The Light FSM enters the On state whenever the button is pressed. The system moves to the Off state 4.5 minutes after the most recent press of the button.

This FSM implementation of the automatic garage door is clearly simpler and more intuitive. The key to achieving this simplicity is to store the direction the door moves not in a state, but in a variable. The opposite of the current value of this variable is output whenever the system leaves the "Moving in direction d state." Also, two different inputs, Button pressed and Limit tripped, both transform "Moving in direction d" state into the Stopped state. The separate Light FSM is similar to our earlier example in **Figure 3.2**, except that repeated presses of the button leave the FSM in the On state—only the passage of 4.5 minutes without a button press moves the system to the Off state.

Any of the implementations we have considered can be made to serve as a model for writing code, but clearly, the FSM with the direction of door motion encoded into a variable is the simplest and least likely to lead to coding errors. In construction of a robot program or any program for that matter, managing complexity is always a vital consideration. Many tools, like the FSM, have the power to aid in this process, but as we have just seen, the magic is not in the tools themselves, but rather in their effective application, a skill acquired through practice.

Beyond introducing finite state machines, the deeper point of this section is: don't follow your favorite method over a cliff. Be always sensitive to the uncomfortable feeling that growing complexity brings on. Excessive complexity is your trigger, indicating that it is time to look for another approach, a different way of analyzing the same system. If you require further evidence of the imperative for simplicity, note that all we have done in this section is to analyze a simple garage door controller—a device with one pushbutton, two limit switches, a one-degree-of-freedom actuator, and one light. Without learning to manage complexity effectively, how can you hope to succeed in programming a multi-sensor, multi-degree of freedom, task-achieving mobile robot?

FSM Example: Escape

Let's now consider a finite state machine example closer to a mobile robotic application—the classic Escape behavior. This is the behavior a robot runs after it has collided with an obstacle. Escape tries to extricate the robot from the near vicinity of the collision.

The robot knows that a collision has occurred because contact with an obstacle compresses the bumper. Many physical mobile robots incorporate a bumper instrumented with two switches, one at either side of the bumper. When the bumper compresses, because of collision with an obstacle, one or both of the "bump switches" trips, alerting the electronics to the collision. Escape is the software routine called on to respond and rescue the robot.

What to do? How the robot should respond to a collision depends on the geometry of the robot. Cylindrically symmetric robots are able to respond effectively to collisions using an especially simple and reliable strategy. The first thing to do is to back up just enough to uncompress the bumper, allowing the robot to turn easily. After backing up, the bumper will no longer be in contact with the obstacle blocking the robot's path and the signal that alerted the software to the collision will vanish—but the robot will remain in a difficult position: stopped lightly touching an obstacle.

The fact that we want the robot to move back a bit before turning, thus eliminating the collision signal, is the reason it is difficult to build a non-ballistic version of the Escape behavior. We could try to make Escape into a servo behavior by having the robot spin right as long as a left bump was present and spin left for the duration of a right bump—but there are problems. What should the robot do when an object presses the bumper directly from the front and both bumper switches are triggered? If the robot backs up, then as soon as the bumper uncompresses, the robot will forget that an object is present and may move forward again, immediately retripping the bumper.

We could solve that problem by never letting the robot back up and instead decide arbitrarily to spin left or right when both bump switches are triggered. But that still leaves us with another problem. If the robot encounters, for example, a wall, then spins until it no longer detects the wall, the robot may stop rotating while touching the wall just lightly enough that the bumper is not triggered. The robot may then resume its forward motion scraping along the wall—perhaps a less-than-effective escape.

To solve these problems, Escape is more commonly implemented as a ballistic behavior. See **Figure 3.6** for an FSM instantiation. Here the robot begins to back up as soon as either the left

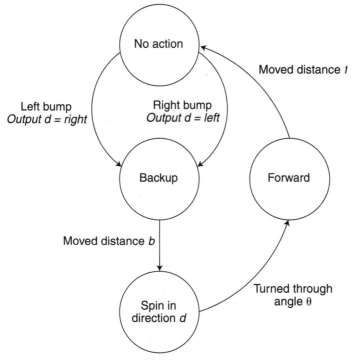

Figure 3.6

One FSM implementation of the Escape behavior is shown in the diagram. In the No action state, the behavior does not compute a command for the robot. If the bumper detects a collision on the left, the No action state outputs d with a value of right and moves to the Backup state. If a collision occurs on the right side of the bumper, the No action state outputs a value of left for the variable d and moves to the Backup state. While in the Backup state, the robot moves backwards until it has retreated a distance b, the system then enters the "Spin in direction d" state. In this state, the robot spins either left or right (as determined by variable d) until the robot has turned a predetermined number of degrees, θ. After the robot has turned θ degrees, the FSM transitions to the Forward state. Here the robot is commanded to move forward until it has traveled a distance f. At that point, the FSM returns to the No action state and no further commands are issued.

or right bump switch closes because of a collision. At that time, a value for the variable d is chosen (the value is output during the state transition). It is through the choice of d that Escape remembers on which side the collision occurred.

If the robot strikes an obstacle head on, it can happen that both the left and right side collision detection circuits trigger at exactly the same time. The FSM in the figure seems not to consider this possibility—how does the system respond if this should occur? Typically, the code that implements the FSM will check one condition before the other. For example, if the possibility of a left collision is tested before the possibility of a right collision, then the robot will decide that a collision has occurred on the left even if a collision on the right happens at the same moment. If it is important to distinguish between left, right, and central collisions, the programmer can include an explicit test for simultaneous left/right collisions and a different outcome can be specified.

Figure 3.7 relates the robot's actions to the states of the FSM. The FSM remains in the Backup state until the robot has moved backward a distance b. When this occurs, the robot transitions to the "Spin in direction d" state and begins to spin in the direc-

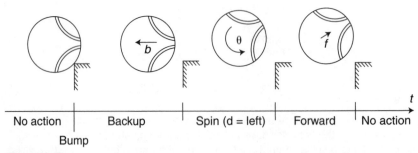

| No action | Backup | Spin (d = left) | Forward | No action |

Bump

Figure 3.7

A robot employing the version of Escape diagrammed in Figure 3.6 might behave as shown upon colliding with an obstacle on the right side of the robot. The time line, t, shows how Escape progresses through the states. Initially, before a collision has occurred, Escape is in the No action state. At the instant a bump is detected, Escape switches to the Backup state and remains there until the robot has moved backward a distance, b. The robot then enters the Spin state, remaining in that state until it has accomplished a spin of θ degrees to the left. Finally, the robot enters the Forward state. It stays in the Forward state until it has moved a distance, f, forward, returning at that time to the No action state.

tion chosen when the initial collision took place. The angle, θ, through which the robot spins, is a parameter of the system chosen to balance competing constraints. If the value is large, say 90° or more, then the robot will generally exit cleanly from walls and other large obstacles. But the robot may fail to find its way into small openings, which impedes its ability to navigate in cluttered environments. If the value of θ is small, the robot will move more easily through clutter but will require many back-and-forth motions when executing a seemingly trivial escape from a wall. The final forward motion through distance f makes sure the robot ends in a position different from where it began the Escape behavior.

FSM Implementation

A version of Escape that is even simpler than the one shown in **Figure 3.7** is diagrammed in **Figure 3.8**. Although crude, this form is frequently sufficient; early implementations of BSim performed Escape in just this way. BSim now relies on the more

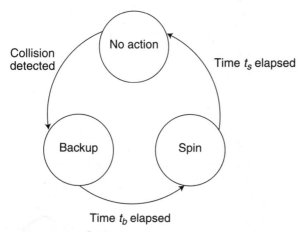

Figure 3.8

This FSM diagram shows the classic Escape behavior in, perhaps, its simplest form. The robot begins in the No action state. When a collision occurs, the system transitions to the Backup state. Here the robot backs up for t_b seconds. When backup is completed, the system enters the Spin state. After t_s seconds of spinning in place, the system returns to the No action state to await the next collision. The values of the parameters t_b and t_s are adjusted experimentally.

sophisticated four-state form. The following BSim Java code implements an Escape behavior close to the one diagrammed in **Figure 3.8**. The key difference between the FSM diagram and the BSim code is that transitions between states are based on elapsed time rather than distance and angle.

```java
/**
 *If the robot is bumping, this behavior performs a set sequence of
 *actions. First it backs up for a set amount of time, and then it spins
 *in place for a set amount of time, then goes forward.
 **/
public void act(Robot robot) {
        int maxWheelSpeed = robot.getMaxWheelSpeed();
        if (state == State.START) {
           if (robot.isBumping()) {
                timeOut.start();
                state = State.BACKUP;
                }
        }
        else if (state == State.BACKUP) {
                robot.setLeftWheelSpeed(-maxWheelSpeed, this);
                robot.setRightWheelSpeed(-maxWheelSpeed, this);
                    if (timeOut.isPast(backupTime)) {
                            timeOut.start();
                            state = State.SPIN;
                    }
            }
        else if (state == State.SPIN) {
                robot.setLeftWheelSpeed(maxWheelSpeed, this);
                robot.setRightWheelSpeed(-maxWheelSpeed, this);
                    if (timeOut.isPast(spinTime)) {
                            state = State.FORWARD;
                    }
            }
        else if (state == State.FORWARD) {
                    robot.setLeftWheelSpeed(maxWheelSpeed);
                    robot.setRightWheelSpeed(maxWheelSpeed);
```

```
        if (robot.isBumping()) {
            timeOut.start();
            state = State.BACKUP;
        }
        else if (timeOut.isPast(forwardTime)) {
            timeOut.start();
            state = State.START;
        }
    }
    else {throw new NullPointerException("Null follow state");
    }
}
```

Much is left out of this example code. We depend, for example, on Java's multitasking environment and make use of variables and constants defined outside the scope of the code. However, you need not understand every statement to gather a basic comprehension of how this code makes the robot behave. There are four states: Start, Backup, Spin, and Forward. As long as no collision occurs, the behavior remains in the Start state where no motor commands are issued. When a collision does happen, the code initiates a timer and sets a new state, the Backup state. On the next call to the Escape routine, the block of code that forms the Backup state is executed. Here the behavior tells the robot to spin both drive wheels backward at the maximum speed. (Of course, additional software stands between the Escape behavior and the motor driver.) When the backup timer expires, the code starts a new timer and switches to the Spin state. The next time the behavior runs, the Spin state code block executes. Here the robot is commanded to spin in place (equal but opposite velocities are sent to the drive wheels). After the spin timer expires, subsequent calls to the Escape behavior are dispatched to the Forward state. Here the robot is commanded to move forward. A collision in this state sends control back to the Backup state. If no collision occurs, Escape returns to the Start state, where it ceases issuing motor commands.

Overloading Behaviors

We saw earlier that ballistic behaviors often cause problems for much the same reasons that deterministic or plan-based robot control schemes suffer. If the situation changes during the ballistic portion of the behavior, the robot may do the wrong thing. What happens if, while the robot is reacting to one collision, another collision occurs? Or perhaps while executing a ballistic response to a just-detected cliff, the robot encounters a wall. Should we add states or special-case code to our behavior to deal with all the possible additional interactions that might occur?

In general, the answer is no. To manage complexity effectively, it is important to avoid overloading behaviors. When behaviors become too complex, we start to lose the advantages of the behavior-based approach. The idea is to write a collection of simple behaviors, each of which manages one specific situation.

Suppose behavior A is designed to handle event 1 and behavior B, event 2. Suppose further that the response of the robot needs to be qualitatively different when event 1 and 2 happen in close proximity from the response required when 1 and 2 occur in isolation. Rather than adding special cases to behavior A to handle event 1 followed closely by event 2, a better solution is to design a third behavior, C. Behavior C responds only when events 1 and 2 occur together and can be optimized for this situation. This choice tends to reduce the overall complexity of the system and leads to more robust performance.

Summary

- A primitive behavior is composed of a control system and a trigger.

- Servo behaviors respond immediately to changes in sensory inputs—when the input changes the output changes.

- Ballistic behaviors, once triggered, follow a predefined path through to completion.

- Finite state analysis is often helpful as a tool for describing, understanding, and implementing behaviors.

- Managing complexity is a necessity in any robot program.

Exercises

Exercise 3.1   The operation of appliances can sometimes be analyzed in terms of triggers and control systems. What is controlled in a clothes dryer? What sensor could be used to trigger the drying operation? (That is, the dryer will shut off when the drying behavior becomes untriggered.)

Exercise 3.2   What triggers an automatic night light?

Exercise 3.3   Sketch out the FSM diagram of a flush toilet.

Exercise 3.4   Does the autopilot of an airplane implement a servo or ballistic behavior?

Exercise 3.5   Does the guidance system of a ballistic missile exhibit servo or a ballistic behavior?

Exercise 3.6   Describe the operation of a clothes washing machine. Sketch the FSM diagram of a washing machine. What parameters are available for user control?

Exercise 3.7   Does the operation of a furnace and thermostat constitute a ballistic or servo behavior?

Exercise 3.8   Use BSim to experiment with the Cruise behavior by implementing a task whose only behavior is Cruise. What happens when you give the left and right drive wheels different velocities?

Exercise 3.9   The automobile's automatic cruise control mechanism is a servo behavior. What events should untrigger this behavior?

Exercise 3.10   Pyroelectric sensors are often used in motion detectors. These sensors produce no output when there is no motion in their field of view but, as a heat source passes in front of the sensor, the voltage output by the sensor goes from zero to a positive value. If the temperature of the object the sensor is focused on is constant, the voltage decays back to zero. When the object passes out of view of the sensor, the voltage becomes negative, then decays again to zero. Describe in general terms how you could use such a sensor to enable a robot to follow a person. What would happen if the robot were placed in a room with an open fireplace?

Exercise 3.11 Build a BSim task using only the Remote behavior. Practice driving the simulated robot using the joystick pad. Add one behavior to this task that will prevent you from driving the robot into a wall.

Exercise 3.12 The garage door opener in the FSM example is actually even more complex than described in the text. Beyond the wall-mounted and mobile pushbuttons that operate the door, the system includes two other sensors. The purpose of both sensors is to avoid injuring pets or people who attempt to move through the opening while the door is closing. An IR break beam sensor is positioned so that it points across the opening at the bottom of the door just above floor level. If the beam is broken while the door is going down, the door stops immediately. A second sensor monitors the current flowing through the door's motor. If, as the door descends, the current goes too high (as it would if an obstacle or person were struck by the door), the door immediately reverses direction and moves all the way back to the fully open position. Draw the finite state diagram of the total system.

Exercise 3.13 The garage door system has one final complication. There is a second button whose only function is to turn on and off the light associated with the door. The light stays on or off indefinitely after this button is pressed unless the door-open/close button is pressed. Then the system ignores the light button and behaves as described earlier. Draw a finite state diagram that includes the light button.

Exercise 3.14 The garage door opener has no reason to turn on the courtesy light during the day. Assume that a light sensor is added to the system to prevent the light from going on automatically during daylight hours. Adapt the FSM diagram to incorporate this new feature.

Exercise 3.15   The Escape behavior diagrammed in **Figure 3.6** doesn't specify what happens if both the left and right bump switches activate simultaneously, as they would if the robot hits an object dead center. What would or should happen in this case? Draw an FSM diagram that incorporates the new functionality.

Exercise 3.16   Suppose that the robot executing the Escape task described in **Figure 3.6** has no rear bumper. Should the robot collide with an obstacle to its rear while backing up, there is no direct way for the robot to know. Suppose further that such a collision forces the drive wheels to stop rotating. Since the robot measures distance by counting wheel rotations, it will never decide that it has moved backward the required distance b. Thus the robot will remain motionless in the Backup state until the batteries run down. Without adding any extra sensors, how could the robot protect itself from such a condition? Write a new FSM that protects the robot from undetected rear collisions.

Exercise 3.17   Does the Cruise behavior, as described in this chapter, constitute an open loop or a closed loop system?

Exercise 3.18   Create an escape behavior that is not blind. Devise a way for the robot to respond to new collisions from any direction during execution of the Escape behavior. Redraw the FSM diagram incorporating your improvements. Can you think of any other ways, possibly involving additional behaviors, that enable the robot to respond to collisions that occur during the execution of a ballistic Escape behavior?

4

Arbitration

In the last chapter, we learned about primitive behaviors. A primitive behavior consists of a control system that makes the robot act in a certain way and a trigger to decide when it is appropriate to take such actions. In order to pursue many goals simultaneously, a robot usually needs many primitive behaviors. As long as only one behavior triggers at a time, things work smoothly. But the roboticist's world is never so comfortable. What ensues when two or more behaviors happen to trigger at the same time, each one wanting the robot to take a different action?

Fixed Priority Arbitration

Imagine a physical robot, similar to the BSim robot, performing the Collection task. The physical embodiment of the collection robot has an additional ability that the BSim robot doesn't need. It can recharge its batteries automatically. Part of the behavior diagram for this robot is shown in **Figure 4.1**. The Wander behavior causes the robot to move about its environment at random. When triggered, the Charge-home behavior issues motor commands that drive the robot toward its recharging station. Charge-home is able to home on the station because an IR beacon marks the station's location.

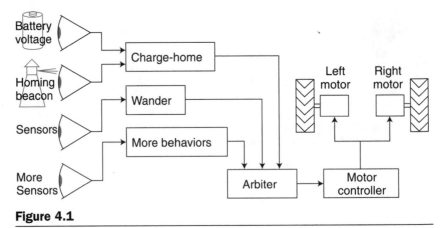

Figure 4.1

In this behavior diagram, Wander, Charge-home, and perhaps several other primitive behaviors compute potentially conflicting commands for the robot's drive motors. Behavior-based systems employ one or more arbiters to resolve the dilemma of which command the robot should obey.

As long as the robot's batteries have sufficient charge, all is well—Wander is the only behavior that issues any commands, so the robot happily roams about. But the Charge-home behavior's trigger monitors the battery voltage. When the voltage reaches a critical level, the trigger trips and Charge-home begins to issue motor commands designed to direct the robot toward the charger. Since Wander randomly tells the robot which way to move, it is possible that there will be no conflict because Charge-home and Wander may issue the same command. But what if they don't? What happens when Wander wants the robot to turn right at the same time that Charge-home issues a turn left command?

There are many possible ways to resolve conflicts such as this, and devices performing different sorts of tasks may choose different resolution schemes. But whatever method we employ, we must be guided by the needs of the system as a whole. The robot's mission is to deliver all the pucks to a position near the light. If the batteries run down before its mission is complete, the robot will stop wherever it is and deliver no more pucks. Thus, when the batteries near exhaustion, it is imperative that the robot head toward the recharging station—if it fails to take this action in this circumstance, the robot will fail to accomplish its task.

Let's consider some possible behavior–conflict resolution strategies. One strategy for resolving a conflict between two competing behaviors is to settle on some compromise between the issued commands. The robot could alternately obey commands from one behavior then the other, or the robot could execute commands from one behavior for a given interval, then switch to obeying the other behavior for a similar time. Yet another strategy might be to average the commands from each behavior. For example, the commands "go forward" from Wander and "spin left" from Charge-home would average to arc left. This might seem "fair" (if we indulge in anthropomorphizing the primitive behaviors), but such a strategy does not guarantee that the robot would reach the charging station before the batteries expire.

Alternately, we might have the behaviors "vote" on which direction to go (in our current example we would need to use the output from one more behavior to break a tie). For example, if two behaviors commanded spin left and one commanded backup, the majority would win and the robot would spin left. Again, looking at the robot from the level of the primitive behaviors, this seems a reasonable strategy for resolving conflicts, but this strategy could not guarantee to get the robot to the charger in time.

The system-level requirement is to keep the batteries charged so that the robot can accomplish its task. The Charge-home behavior is designed to produce the correct response when battery power is low. Any conflict resolution strategy that does not pass commands from the Charge-home behavior to the motors or any strategy that dilutes these commands ultimately degrades the robot's ability to perform its task.

There is a very simple strategy to ensure that the most appropriate commands are always followed. Since we know in advance that commands issued by Charge-home are more urgent than commands from Wander, we can give commands from Charge-home permanently higher priority than commands from Wander. When both behaviors send commands to the arbiter, the higher-priority commands from Charge-home are passed through to the robot's motors, while the lower-priority commands from Wander are ignored.

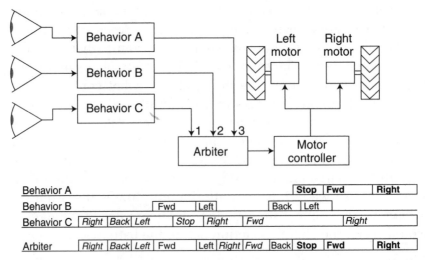

Behavior A						**Stop**	Fwd	**Right**
Behavior B			Fwd	Left		Back	Left	
Behavior C	Right	Back	Left	Stop	Right	Fwd		Right
Arbiter	Right	Back	Left	Fwd	Left	Right	Fwd	Back	**Stop**	Fwd	**Right**

Figure 4.2

In a fixed-priority arbitration scheme, multiple behaviors send their commands to an arbiter. The programmer decides the priority of each behavior in advance and no two behaviors are allowed to have the same priority. When commands from different behaviors conflict, the highest-priority behavior wins and only that behavior's commands are passed on to the robot's motors (or other scarce robot resource). The lower graph illustrates what happens when commands of various priorities conflict—only the highest-priority commands emerge from the arbiter. If no commands are sent to the arbiter by any behavior, the arbiter outputs a default command, usually stop.

The arbiter in this case uses a *fixed-priority* scheme to decide the winner. Fixed priority means that the programmer has decided in advance which behavior ought to win any time a conflict occurs. Each connection from a behavior to the arbiter is given a unique priority. In **Figure 4.2** Behavior A has priority 3, Behavior B has priority 2, and Behavior C has priority 1. Larger numbers indicate higher priorities corresponding to greater urgency.

If Behavior C is the only behavior sending a command to the actuators, then the arbiter delivers that command. However, if Behavior B decides that it too wants to send a command, the arbiter passes Behavior B's command and discards the command from Behavior C. If Behavior A sends a command, its command takes priority over either B or C and only Behavior A's command reaches the actuators.

Figure 4.3 contains the behavior diagram of the Collection task we saw in the first chapter. As used in this text, a behavior diagram includes a representation of the primitive behaviors that implement a task and the sensory systems that the robot possesses. The diagram also shows how information flows from one component to another. By convention, the relative priority of the behaviors connected to a particular arbiter will be indicated by the position of the source behavior. In the figure, Cruise is lowest on the page and has the lowest priority of all the behaviors connected to the arbiter; Escape has the highest position and consequently the highest priority. The arbiter decides which of the competing commands to pass along to the motor controller. The motor controller adjusts current and voltage to the motors in such a way as to carry out the single command that it receives.

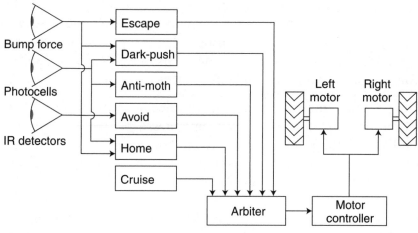

Figure 4.3

The arbitration scheme used by the Collection task is shown. By convention (at least in simpler behavior diagrams), primitive behaviors higher on the page have higher priority. Thus, numbering the inputs to the arbiter is not usually necessary. Here Escape is the highest-priority behavior—its commands reach the motors whenever it is triggered. Cruise is the lowest-priority behaviors—its commands are executed only if no other behavior is triggered.

When to Arbitrate

Under what circumstances is arbitration needed? Any time there is competition among behaviors for scarce resources, arbitration is typically necessary. A robotic resource is scarce if it is possible for the demand for that resource to exceed the supply. The robot's drive motors are a necessarily scarce resource—in the Collection task example, all six primitive behaviors can demand control of the robot's motors, but there is only one pair of motors available to satisfy that demand. The BSim robot has only one arbiter, but in general, it can easily happen that a robot has several resources that require arbitration. For example, a mobile platform that also has a manipulator for picking up objects could have one arbiter for managing mobility system commands and a second arbiter for managing manipulator commands.

It is even possible to have an arbiter manage more abstract "resources." Competing behaviors might vie to control a certain robotic function, for example, a complex sensory system. The parameters of such a system might need to be adjusted and the attention of the system might need to be focused on a single point interesting to the controlling behavior. Arbitration can be applied to smaller grained elements as well (see the section "Subsumption Architecture" later in this chapter).

Some robots include a speaker or piezoelectric buzzer. The buzzer may play tones or annunciate messages that indicate the condition of the robot. Since only one message can be meaningfully played at a time, it may be useful to have an arbiter manage commands destined for the buzzer. Behaviors need to send commands only to relevant arbiters. Just because the commands from a behavior emerge victorious from one arbiter does not necessarily mean that that behavior will be the winner at other arbiters to which it is connected. The various arbiters can establish different priorities for the same behavior. A complex example is shown in **Figure 4.4**. But as in all robot design, the needs of the system as a whole are paramount. Always settle first on the exact global behavior you want to emerge when connecting behaviors to arbiters and assigning priority.

In general, behaviors operate with no knowledge of the state or status of any other part of the system except the sensors. However, sometimes it is necessary for a behavior that wants to control the robot's motors or other resource to know if control

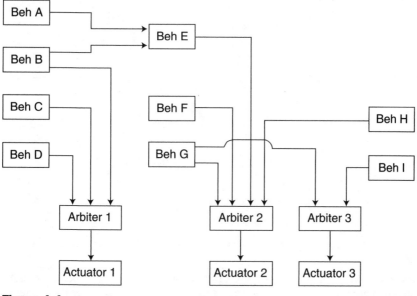

Figure 4.4

A behavior control system can be arbitrarily complex. Behaviors can compute outputs for other behaviors. There can be multiple arbiters; the various behaviors can compute commands for some arbiters and not others. The commands from one behavior can be assigned one priority at one arbiter and a higher or lower priority at another arbiter.

has been granted. Suppose, for example, that an escape behavior has triggered. Escape wants to move the robot backwards, away from whatever caused the collision. To do this, Escape needs to know if it has control before it starts counting the distance the robot has backed up. If Escape is suddenly granted control partway through its ballistic action, the robot would not react as desired. Maybe, for example, the robot would then spin without having first backed up or move forward without having spun. Or a behavior that controls the piezoelectric buzzer may need to know when it has been granted control so that the behavior can play a tune from the beginning rather than picking up somewhere in the middle.

To accommodate behaviors' needs for control information, arbiters are often designed with an output that indicates the winner of the arbitration. See **Figure 4.5** for an example. Behaviors can use this output from the arbiter exactly as they would use

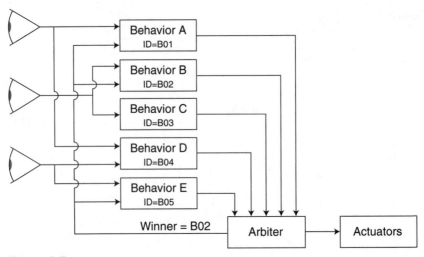

Figure 4.5

Behaviors operate with no knowledge of the internal state of other behaviors but some behaviors do need to know, having requested control, whether the arbiter has granted control. As in this diagram, arbiters are sometimes constructed to include an output line, available to all behaviors, that reports this information. One way to implement this feature is to assign a unique symbol or identifier to each behavior. The arbiter outputs the identifier of the behavior given control on its "winner" line. Each behavior can then compare its own identifier with the value on the output line to determine if it is in control of the arbitrated resource. Here, Behaviors A, B, and E need the arbitration-winner information and Behavior B is the current winner.

any other sensory input. Along with the motor (or other resource) commands each behavior sends to the arbiter is a symbol unique to that behavior. If the control-desiring behavior finds its own symbol on the arbiter's output line, then that behavior can determine that it has won the arbitration and is in control.

To maintain a "clean" system, one that is easy to modify, expand, and debug, it is essential that no behavior be allowed directly to modify another behavior's internal state. Although it is sometimes tempting to use local information, created and maintained by one behavior, as a variable in computations made by another behavior, it is always a mistake to do so. Such behind-the-scenes actions break the behavior-based paradigm and begin the slide toward organically structured programs. If it seems that there is simply no way to construct one behavior without using local information from another behavior, then it is likely that you have chosen an awkward structure for your program. In this case, you should either make the internal information from one behavior an explicit output available to all other behaviors or you should rewrite both behaviors combining the two into one.

Graceful Degradation

Things never go smoothly for robots operating in the real world. In particular:

- A command intended to direct the robot to move in a particular way instead, because of uncontrollable environmental effects, causes the robot to move in a different way.

- The robot's program makes an assumption about the world that turns out not to be true.

- The robot's sensors fail outright or produce false negative or false positive results.

(A false negative means that the sensor did not react when it should have. A false positive occurs when a sensor reports a condition that does not exist.)

Regardless of all this, we want our robots to soldier on. Even when things don't go exactly as planned, we would like our robots to keep functioning. Performance will necessarily suffer when important information goes missing or when good motion commands go bad but, rather than break down catastrophically, the robot program ought to do as well as is possible under the circumstances. The ability of a system to continue to perform at a reduced level in the presence of subsystem errors and failures is known as *graceful degradation*.

The behavior/arbitration structure we have been discussing lends itself seamlessly to graceful degradation. In the example shown in **Figure 4.6**, many behaviors, along with their associated sensors, attempt to keep the robot moving toward its goal (as decided by the Navigate behavior). The Sonar-avoid behavior is

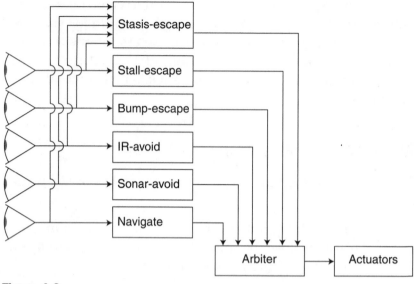

Figure 4.6

This behavior diagram illustrates a navigating robot that is especially adept at dealing with sensor problems. Here five different hazard response behaviors can take over from the Navigate behavior. This can keep the robot moving even when sensors fail to report conditions the programmer expects the sensor to detect (see text).

provided to monitor the long-range sonar sensor and to steer the robot generally along extended open paths. While still some distance off, the robot will turn away from sonar-detected obstacles. But sonar sensors are easily fooled. Smooth surfaces struck at shallow angles reflect sonar beams forward, not back toward the robot—thus no echo is ever detected, causing the sonar system to report a false negative. When this happens, the robot believes that the path is clear even if it is not.

In this case, the Sonar-avoid behavior fails to trigger. But as the robot approaches the acoustically specular[1] object, typically the IR obstacle detectors will sense an obstacle and drive the robot

---

[1]An acoustically specular object is one that behaves as if it were a mirror for sound waves—the waves bounce off in one direction. We would prefer to deal with acoustically diffuse objects where the sound waves bounce off in all directions. In the latter case, some of the sound energy is bound to bounce back to the detector.

away. When the sonar system is fooled, robot performance suffers a bit and the robot no longer drives quite so smoothly away from obstacles; even so, the robot continues to operate. But perhaps along with being too smooth and set at too oblique an angle, the obstacle's surface is also too dark to reflect IR radiation reliably back to the IR obstacle detector. The IR detector also fails to sense the object, thus generating a false negative of its own.

With no signal from the IR detector, the IR-avoid behavior does not trigger and the robot collides with the obstacle. The robot bumper now compresses and triggers the Bump-escape behavior. Having failed to avoid the obstacle, the robot must now back up and try to turn away. Typically, bump sensors can detect, at least crudely, where a collision has occurred, say, on the right or the left of the robot. This knowledge can give the robot an indication of how to respond. Once again, system performance falls a notch. The robot is forced to retrace some of its path and to escape a collision, but still it turns away from objects as well as possible. The robot continues to perform its navigation mission but less smoothly, and its overall progress slows.

But suppose there is yet another difficulty, say, that the bumper is not a full-coverage bumper or that it has a dead spot at some point and it is at exactly that point that the troublesome smooth, dark, oblique object contacts the bumper. Now the bumper fails to report the collision and accordingly, the Bump-escape behavior never triggers. We are left with our robot pressing against the object with all its might. Fortunately, an over-current sensor is available to detect this condition. When the drive motors are asked to work too hard, as can happen, then they force the robot against an object, and motor current goes up. Too high a current for too long a time is sensed and is used to trigger a Stall-escape behavior.

Stall-escape knows only that whatever the drive motors are doing is not having the desired effect. Presumably robot motion is blocked by something. Although there is no reliable way to tell where the blockage is located with respect to the robot, Stall-escape can at least command the robot to retreat and spin. There is a good chance that the robot will break away from whatever has halted it if the robot takes this action.

Again, robot performance is further degraded. Now the robot can't even determine why it is unable to move or where the feature-preventing motion is located. But the robot still has an effective response. It may move toward its goal slowly and haltingly, yet it proceeds.

Let's imagine one final trial for our long-suffering robot. The robot happens to be driving on a freshly polished floor when it encounters the demon obstacle. Because the floor is quite slick, the wheels start to spin when the robot becomes blocked. Freely spinning wheels do not cause the drive motors to consume excessive current, thus, the over-current sensor does not measure a current higher than the stall threshold and Stall-escape never triggers.

A stasis sensor is a sensor that trips when something fails to change. Such a sensor might use a vision system to detect the absence of motion or a stasis sensor might monitor the rotation of a passive caster wheel (tripping when the wheel fails to turn). It is also common to program virtual stasis sensors. If the output of all the other sensors the robot possesses fails to change for some length of time, then the virtual stasis sensor concludes that the robot is not in motion. The tripping of the stasis sensor triggers the Stasis-escape behavior, causing the robot to execute a sort of panic procedure. The robot may spin and back, possibly randomly, in an attempt to free itself from this most generic of faults. With Stasis-escape in control, performance has declined to the lowest possible notch but even under these circumstances, given enough time, the robot may be able to accomplish its task, even in an environment filled with the sort of uncooperative obstacles postulated in this example.

We have assumed that a difficult-to-detect obstacle and a too-slick floor are responsible for all the (temporary) trouble in this example. But note that even if the cause of the problems had been permanently inoperative sensors, we would have the same result—in all cases, the robot would have done the best it could possibly do under the circumstances. (This is true of permanent sensor failures only if the sensors fail by effectively generating false negatives; constant false positives require more sophisticated programming—see below.) Note also that our behavior/arbitration structure lets us achieve this helpful condition with no effort. We wrote no code that explicitly instructed the robot to determine if a sensor was working properly. Remarkably, graceful degradation is a property that emerges from the overall structure of the system.

Although this example contains no explicit code to deal with sensor failure, it is always important to contemplate the ways in which sensors can fail. It is best practice to assume that every sensor will spend a good fraction of its time reporting inaccurate results. Consider whether, under these circumstances, robot performance will degrade gracefully or if the transmission of a single false reading or a single noise glitch will bring your robot task to an embarrassing, twitching halt.

If it seems far-fetched that so many sensors could fail to do the job for which they were intended, think again. In the real world, difficulties with sensors and the environment are very much the norm rather than the exception.

Figure 4.6 offers a case study of only one type of graceful degradation; there are many other types. It is important to ensure that a degradation path exists for any critical sensor. For example, a robot operating in an area that contains descending stairs needs a cliff sensor that enables the robot to detect the stairs and turn away so that the robot can continue its task. We would not want the robot to become a safety hazard should the cliff sensor generate a false negative. Thus, this sensor should be backed up by some other system. Perhaps the chassis of the robot is arranged in such a way that if any of the robot's wheels drive over an edge, a physical stop contacts the floor in such a way that the robot is prevented from moving further.

The Path Not Taken

The example of the previous section illuminates a crucial difference between the behavior-based approach to robot design and other approaches. Note in particular what we did *not* do. Rather than constructing a robot with redundant modes of operation, the more typical design and testing scenario goes like this:

A team wants to build a robot able to navigate to a particular point while avoiding obstacles. The designers observe that sonar sensors are usually able to detect obstacles while the obstacles are a good distance away from the robot; therefore, a navigation sensor and an inexpensive sonar ranger are the only sensors the robot really needs. The team feels that this choice will minimize robot cost and design time. The robot is built and a program is written that avoids objects based on sonar range information. The team makes a few test runs with the robot and it becomes clear that the reasoning is correct—usually the sonar system does detect obstacles. But it's also clear that "usually" isn't good enough. Whenever the robot happens to encounter an undetectable obstacle, it becomes hung up, unable to complete its task—and the robot seems especially adept at locating obstacles the sonar system can't see.

What to do? Clearly the program works when it gets good data from the sonar system. So the problem can't be in the software; it must be that the sonar system just isn't up to snuff. The team makes a decision to switch to a more expensive obstacle-detecting system that uses multiple, higher-frequency sonar sensors. Higher frequency means shorter wavelength; shorter wavelength means that smaller objects can scatter the ultrasonic waves and therefore fewer obstacles go undetected.

The refitted robot does work better. It can handle more marginal obstacles and can detect those obstacles from a larger set of approach angles. But there are still too many situations that trap the robot; the robot cannot reliably complete its task in the environments where it is required to operate. Again, the program and strategy appear sound—the problem, the team concludes, is the quality of the data.

Maybe the robot's difficulty stems from choosing a sonar system in the first place. Perhaps a much more capable scanning laser ranging system would be more appropriate. Such systems are available at a cost of a few thousand dollars. The robot may have to be redesigned and made larger to accommodate the laser device and the laser scanner's data processing requirements now exceed the abilities of the robot's original 8-bit microprocessor—that will have to change too. Rebuilding the robot with a larger processor will add even more to the cost. At some point, the project becomes so expensive that the team is forced to make a painful decision—either abandon the project altogether or go forward with the existing equipment. The latter choice means that to use the robot, an operator must cleanse the environment of any obstacles the robot can't see before starting a run.

At this point, an honest assessment of the project would conclude that the team has not built a practical robot—it has instead built a high-tech laboratory demonstration, creating a learning opportunity, not a generally useful product.

The notion of graceful degradation facilitated by a behavior-based approach to robot design allows us to dodge such woeful tales. In behavior-based robotics, we do not employ a single expensive sensor from which we must demand unattainable levels of precision and reliability. Rather, we achieve superior results using a combination of relatively unreliable systems that work together to deliver robust performance.

Sensor Qualification

Dust, dirt, vibration, and age can all affect sensor operation—and never in a helpful way. To maximize reliability, a robot ought to avoid depending on any sensor that can be shown not to work as specified. As mentioned above, it is usually relatively easy to ensure that our system degrades gracefully in the presence of sensor failures if the failure takes the form of a false negative. But what if the sensor produces a false positive? A sensor that is "stuck on" may constantly indicate a hazardous condition. This will cause the robot constantly and improperly to respond to phantom peril; it may be prevented from completing its task. Is there a way to degrade gracefully in the case of false positives?

See the behavior diagram in **Figure 4.7**. This diagram includes a sensor-qualifying behavior whose purpose is to verify the functionality of a sensor before the robot acts on the output of that sensor. Consider, for example, an IR proximity sensor, which works by emitting a beam of modulated IR radiation. The sensor reports that an obstacle has been detected when the associated receiver detects radiation reflected from an object. However, such receivers are very sensitive and it sometimes happens that IR radiation from the emitter finds a direct path to the receiver without bouncing off an obstacle. This might occur if the emitter window were scratched in such a way that it scattered light

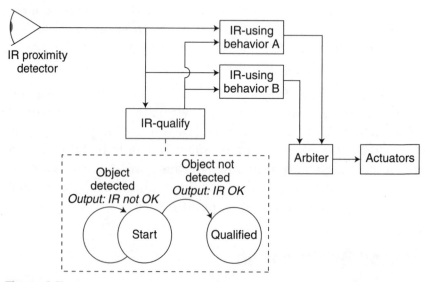

Figure 4.7

A sensor-qualifying behavior can be implemented to protect the robot from sensors that fail by generating false positives. In general, such qualifying behaviors operate by assuming that a sensor is non-functional until it has been observed exhibiting all of its relevant states. Here the IR-qualify behavior determines when the IR sensor functions properly. Inside the IR-qualify behavior is an FSM. The FSM begins at power up in the Start state. As long as the IR proximity detector senses a nearby object, the FSM remains in the Start state. As soon as the detector fails to sense an object, the FSM moves to the Qualified state and remains there for the duration of the run. The IR-qualify behavior informs other behaviors when it is safe to use IR data.

toward the receiver or if a bit of debris had become stuck to the robot near the emitter. In these cases, the sensor would constantly report an obstacle present regardless of whether there were one.

Qualifying information can be wired into the control structure in a couple of ways; one method is shown in **Figure 4.7**. There a connection is made from the qualifying behavior to each behavior that might be adversely affected by false positive indications. Each such behavior can be held in an initial state, issuing no motor commands, until the sensor-qualifying behavior has verified that the sensor is functioning. Another approach is to build a virtual sensor behavior, as in **Figure 4.8**. The virtual sensor behavior takes input from a real sensor, but does not pass on positive indications until sensor functionality has been verified.

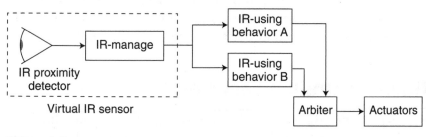

Figure 4.8

Building a virtual sensor provides a method for eliminating false positives that is potentially less complex than wiring the qualifying information into the control structure. Here a virtual IR proximity detector, one that does not generate false positives, has been constructed from a physical sensor and the IR-manage behavior. IR-manage does not pass positive indications of obstacles until it has observed the sensor in both its detecting and non-detecting states.

In many cases, as in our two examples, actual verification can be quite straightforward. Suppose that when the robot powers up, the sensor reports the presence of an object. But is an object actually present or has the sensor failed with a false positive indication? If, as the robot moves, the obstacle indication vanishes, then we can conclude that the robot simply started up near an object. As soon as the robot moved away, the sensor stopped seeing the object and accurately reported this result. Having seen the sensor in its no-obstacle state, we know that the sensor is not constantly producing false positive indications and we can accept that the sensor is functional.

In most cases, a hobby robot can dispense with sensor-qualifying behaviors. If the robot misbehaves, the roboticist will observe this fact and take action to remedy the cause. But for a robot that becomes a commercial product, such care is essential. Consumers and business customers will expect their robots to function as well as they can for as long as they can. Adding sensor qualifying software generally costs nothing (except for the development time) and can greatly improve the robustness of the robot.

Even an autonomous hobby robot might benefit from sensor qualification if there is a possibility that a sensor could fail during a long contest when the robot is beyond the control of the roboticist. In that case, sensor qualification could make the difference between degraded performance and immediate defeat.

Other Arbitration Schemes

Fixed-priority arbitration is by no means the only method for combining conflicting outputs from multiple behaviors. However, while novice robot programmers often seem to expect that their particular application will require some exotic form of arbitration, this rarely turns out to be the case. I have had to resort to another more complex form of arbitration on only one occasion over the course of about 15 years. Do not begin by assuming that your project is too complex to benefit from fixed-priority arbitration. Other forms of arbitration, especially *ad hoc* forms, can quickly cause your code to bloat to project-threatening levels of complexity.

Variable Priority

Pondering fixed-priority arbitration schemes invites the thought, "Fixed-priority arbitration must be just a subset of variable-priority arbitration; therefore, maybe variable priority is better."

What if we build an arbiter where the priorities assigned to the various behaviors can change during run time?

Such schemes are workable, but in robotics, they are only rarely worth the trouble. In choosing to build such a system, you will have to deal with issues such as: What determines the priority ordering? How do you ensure that two different behaviors never have the same priority? Are there any conditions that will lead to rapid, cyclic priority reordering (this may stop the robot from making progress)?

An important practical disadvantage of any variable-priority scheme is that robot behavior cannot be easily understood simply by examining the behavior diagram. Because the way a variable-priority system works can shift from moment to moment, exactly what the robot is doing and why can be challenging to determine. This makes debugging tricky. With a well-reasoned decomposition of the problem, a fixed-priority scheme can almost always be engineered to accomplish a given task.

Subsumption Architecture

A paper[2] by Rodney Brooks in 1986 marked the formal beginning of behavior-based programming. In the paper, Brooks describes a scheme called subsumption architecture. Subsumption was inspired by the development of brains over the course of evolution. In this process, the lower, more primitive functions of the brain are never lost; rather, higher functions are added to what is already there. At the core of every human brain is the remnant of an earlier reptilian brain.

Likewise, in subsumption, the most primitive abilities of the robot are programmed first. Things like motor control, behaviors responding to collisions, behaviors responding to the presence of nearby obstacles, and so on, are programmed and wired together. Next, to expand and improve on the capabilities of the robot, new, more sophisticated behaviors are added on top of the

---

[2]See "A Robust Layered Control System for a Mobile Robot," Rodney A. Brooks. *IEEE Journal of Robotics and Automation.* RA-2, 14-23 April 1986.

existing behaviors. When necessary, the functionality of the old behaviors is subsumed by the new behaviors. But the older behaviors are always present, ready to control the robot whenever the higher levels of behavior are not appropriate.

Given this conceptual approach, it is important to be able to break into the wiring established by earlier behaviors and insert commands from later, higher-level behaviors. **Figure 4.9** shows how this works. Fixed-priority arbiters (with two inputs each) are scattered throughout the code. In the figure, each circle labeled with an S or I is a type of arbiter. The way to think of a subsumption construction is that the robot begins with, perhaps, one behavior connected to the actuators. Then a new behavior is added on top of the first one, the original wire is broken, and a new signal is injected from the new behavior. The new signal (marked by the arrowhead) always has priority over the old.

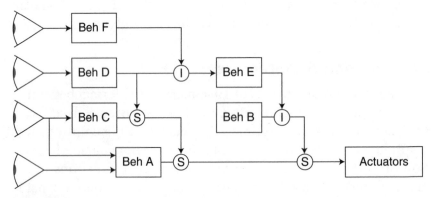

Figure 4.9

Subsumption architecture uses multiple arbiters. Arbitration occurs in structures called suppressor or inhibiter nodes, indicated by S and I, respectively. Here we might imagine that initially behavior A was developed and debugged. Then behavior Beh B was added. To incorporate Beh B, the wire connecting Beh A to the Actuators was broken and the signal from Beh B inserted into the wire through the suppressor note. Development would proceed in this manner, constructing and debugging low-level behaviors, then adding higher-level behaviors. The robot works at every stage, its functionality increasing as more behaviors are added.

The S- and I-labeled circles are called suppressor and inhibitor nodes, respectively. A suppressor node is most similar to the fixed-priority arbiter described earlier. Such a node has two inputs and one output. The input marked with an arrowhead always has the higher priority so that when signals on the two inputs collide, only the signal traveling along the arrowhead input appears at the output. The higher-priority input thus suppresses the lower-priority input. The lower-priority input appears at the output only if there is no signal on the higher-priority line. The multi-input arbiters we have seen before can be constructed using a collection of suppressor nodes. See **Figure 4.10**.

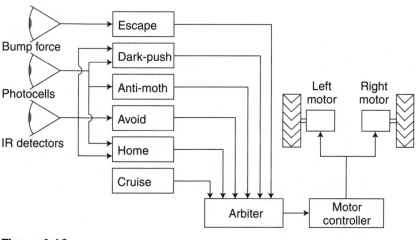

Figure 4.10

Here the Collection task is rewritten using the subsumption architecture formalism. The familiar multi-input arbiter is replaced with several ganged suppressor nodes. Written either way, system function is the same.

A type of arbiter we have not previously discussed is the inhibitor node. An inhibitor node makes it possible for one behavior effectively to turn off another behavior by sending a message. Like a suppressor, an inhibitor node has inputs with different fixed priorities. But in an inhibitor node, the signal on the higher-priority input does not replace the lower-priority input at the output. Rather, when a higher-priority input is present, no signal is output from the node at all. That is, the higher-priority signal inhibits the lower-priority signal, preventing the inhibited behavior from sending its message—just as if the lower-priority behavior had not tried to send a message at all.

So far, what it means for commands from two or more behaviors to collide, to request control at the same time, has not been clearly stated. When a behavior wants control, does it send a single message containing its control commands? Or does the behavior resend its control message as often as the system will allow? Or does the behavior perhaps set a flag on some unseen control line stating that control is requested? In the original implementation of subsumption architecture, when a behavior wanted to get and keep control, it had to send messages as often as it could. The behavior had to do this because of the time constant, t_n, associated with each node. After a higher-priority message had been channeled through a node, the node would refuse to pass a message from the lower-priority input until a time, t_n had passed. The time constants were adjustable on a per-node basis.[3]

Subsumption architecture may be a somewhat more general scheme than the fixed-priority system developed above. However, the subsumption representation is more complex and, given variable time constants at each node, it is not always possible to understand what the system will do simply by looking at the behavior diagram. I have not found inhibitor notes or variable time-constant nodes essential. In the rare cases where a construction like an inhibitor node or a node with a variable time constant might be useful, such functionality can easily be implemented at the behavior level, rather than the arbiter level.

To answer the message-collision question raised: Many simple, fixed-priority arbitration schemes use an unseen control line associated with every command connection. When a behavior wants control, it sets a flag on that control line. The arbiter copies whatever command is present at the output of the highest-priority behavior, whose flag is set, to the output of the

---

[3]Beyond writing a paper describing the ideas behind a layered control system for robots (subsumption architecture), Brooks also developed a versatile software system to support his subsumption ideas. The system is called the Behavior Language and you can find a full description in: *The Behavior Language; User's Guide*, A.I. Memo 1227, April 1990. (A publication of the MIT AI Lab, http://www.ai.mit.edu/research/publications/publications.shtml.) Behavior language is suitable for even the least powerful 8-bit microprocessors and has been instantiated in several such processors.

arbiter. Thus, no explicit time constant is associated with arbitration and behaviors need not send their commands repeatedly; once will do. A winning command remains in effect until the issuing behavior changes the command, until the behavior clears its control flag, or until a higher-priority behavior takes control.

Another issue broached by subsumption architecture is, what exactly does it mean to "connect" one behavior to another? Is a connection really just a global variable to which both behavior and arbiter (or behavior and behavior) have access? Or is it something more exotic and mysterious? The original subsumption implementation connected behaviors using explicit constructions called wires—wires had sources and destinations.[4] In general, in behavior-based robot control schemes, connections can be implemented in any number of ways. The key idea is that the transfer of information between behaviors must be explicit—unless you establish a connection with a mechanism that can be represented by a line on the behavior diagram, no connection should exist. Only explicit sources of the information can change the information, and only explicit destinations can access the information. Connections can be implemented as global variables (if the programmer can resist the temptation to misuse them), but it is strongly preferable to hide the details of the connection and implement and access connections using only special-purpose methods.[5] In Chapter 8, we will examine at some depth one straightforward macro-based method for implementing connections.

Motor Schema

At least for certain types of problems, a method known as motor schema[6] offers a more inclusive way to combine conflicting

---

[4]Indeed at least one implementation of subsumption made connections with actual physical wires. See *A Colony Architecture for an Artificial Creature* by Jonathan Connell, AI-TR 1151, Cambridge, MA, MIT, 1989.

[5]Hiding the details of your program may sound like a bad idea, but in fact it is a common and helpful practice. For example, the complier mercifully conceals the messy details when a statement in a high-level language is converted into assembly language instructions.

[6]See "Motor Schema Based Navigation for a Mobile Robot: An Approach to Programming by Behavior" by Ronald Arkin. *Proceedings of the IEEE Conference on Robotics and Automation*, Raleigh, NC, 1987, pp. 265–71.

behavior outputs than by simply picking a winner. See **Figure 4.11**. Here we imagine two behaviors—a homing behavior whose purpose is to drive the robot toward a goal and an avoid behavior that tries to drive the robot away from every obstacle. At any point in space, the command that each behavior individually computes for the robot is clear. As shown in part (b), the homing behavior commands a motion directly toward the goal as the vectors indicate. Similarly, the avoid behavior commands the robot to move directly away from the vicinity of any object it encounters, as shown in part (c).

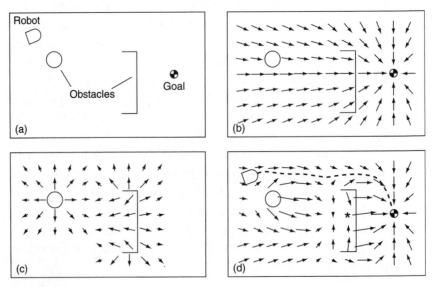

Figure 4.11

Motor schema implement arbitration for certain sorts of tasks using vector fields. In part (a), a robot positioned as shown is to make its way to the goal while avoiding the obstacles. One well-known approach to solving such problems employs potential-generated vector fields. The goal possesses an attractive potential that everywhere draws the robot toward the goal. Part (b) illustrates the vector field generated by such a potential. The length and direction of an arrow indicate the strength and direction of the force the robot feels at the position of the arrow. Obstacles possess a repulsive potential whose magnitude decreases with distance from the obstacle, as shown in part (c). Adding the vector fields of (b) and (c) yields (d). (The asterisk marks what is known as a local minimum.) All the robot need do to travel from the start position to the goal is to move in the direction of the local vector. At any given position following the vector produces the best compromise between progressing toward the goal and avoiding collisions. The dashed line shows the robot's path.

Rather than simply picking a winner, Home or Avoid, we can in this case compromise on the direction the robot moves. Part (d) shows the vector sum of the attractive and repulsive vector fields. At every point in space, the robot computes the best compromise between homing and avoiding. The robot can then smoothly follow a path from start to goal, avoiding all objects along the way.

When it is possible to employ them, methods such as this can be quite useful. This and analogous methods allow us to treat the various constraints on the robot in the same way. Homing and avoiding, rather than being different behaviors, are both just potentials with different values at different places. Casting all constraints in a common language can make possible the discovery of optimal solutions. Unfortunately, at least a couple of problems remain. First, we rarely get to know the global vector field as so beguilingly shown in the figure. Instead, the robot uses its sensors to gather information, and from this information the robot attempts to construct the locally relevant portions of the global field. But the robot may be unable to do this consistently.

The second problem is that we haven't actually incorporated *all* the constraints we need to move from start to goal avoiding collisions. Motor schema makes use of a potential method and all potential methods struggle with a phenomenon called *local minima*.[7] A local minimum is a point in the summed vector fields where the robot can become trapped. The starting point in our example was fortuitous; had the robot began its journey at the center left rather than the upper left, it would have traveled into the (non-convex) bracket-shaped object and stopped. The stopping point, marked by an asterisk in **Figure 4.11**(d), is a point toward which all nearby vectors are directed.

---

[7]The minimum referred to is a local minimum of the potential field. We generate the total scalar potential, $p(x,y)$, by adding together the potential of the goal (the closer to the goal, the lower the potential) and the potential of the obstacles (the closer to the obstacle, the higher the potential). The vector field is just the gradient of the potential, $\nabla p(x,y)$. Non-convex objects or arrangements of objects of any shape can easily create potential field minima, points where attractive and repulsive forces are balanced. Should the robot wander into such a point, it can become immobile because nothing that the robot is able to sense locally can tell it which way to move in order to escape.

To escape a local minimum, the robot must do something other than simply follow the vector field (following the vector field was the strategy that got the robot trapped in the first place). A common tactic to escape such points of no progress is to have the robot alter the potential—increase the potential of previously visited places until the troublesome spot is no longer a minimum. This will work *if* the robot has accurate positioning information. Another tactic is to begin wall following or switch to some other strategy when progress toward the goal halts.

Least Commitment Arbitration

Sometimes behaviors overstep their authority. In all our fixed-priority arbitration examples, we have seen cases where, depending on the circumstances, one behavior or another was triggered and that single behavior assumed complete control of the robot. But it can happen that a single behavior does not know what the robot *should* do, but rather only knows what the robot should *not* do.

Consider the example shown in **Figure 4.12**. The robot in this figure uses a synchro drive system (see Appendix A) to acquire capabilities beyond those of the robots we have considered up to this point. A synchro drive robot can move in an arbitrary direction without first spinning to face in that direction. Suppose that the robot's task involves moving as close as possible to obstacles without actually colliding with them. (A cleaning robot attempting to sweep debris from the edges of a room might need to do this.) For this purpose, the robot comes equipped with a full-coverage proximity detector (a capacitive sensor might perform this function if the conductive and dielectric properties of objects in the robot's environment are similar).

To specify motion for the synchro drive robot completely, a behavior cannot simply output the speeds of two wheels. Rather, the behavior must specify both a direction and a rotation rate for the robot (e.g., move diagonally to the right while rotating left). This is more information than we have previously required our behaviors to supply. Now consider the situation of an avoid behavior monitoring the situation shown in the figure. The front left corner of the robot is about to strike an object. According to

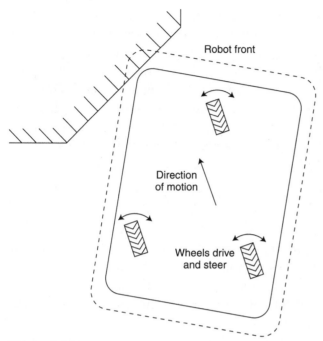

Robot front

Direction of motion

Wheels drive and steer

Figure 4.12

Here a rectangular robot equipped with a synchro drive (the robot can move in any direction without first spinning to point in that direction) encounters an obstacle. The robot is protected by a full-coverage proximity-sensing system indicated by the dashed line. Able to move or rotate in any direction, this robot has more instantaneous degrees of freedom than other robots we have considered. In order to pick the best motion to make, it may be helpful to use a least-commitment arbitration scheme.

our earlier examples, we might have the avoid behavior assume control and drive the robot away from the hazard.

But which way should the robot move? Should it back up? Spin to the right? Move diagonally right? The avoid behavior in this case does not actually know the best direction to move the robot; all Avoid knows with certainty is the directions in which the robot should *not* move. The robot should not move forward, left, or diagonally left, and it should not rotate to the left. But any other motion will tend to move the robot away from the collision hazard and would be an acceptable motion from Avoid's point of view.

We could, in this case, simply have the Avoid behavior choose a direction/rotation arbitrarily, take control of the robot, and move in that way. But arbitrary decisions can turn out arbitrarily badly. It is always best to avoid such decisions because often an arbitrary choice may turn out to be unhelpful for reasons not known to the decision maker. Rather than have Avoid commit the robot to a possibly bad move, a superior strategy is to make the least commitment possible. Perhaps we can somehow have all the hazard-avoidance behaviors identify motions that would be dangerous in the given situation, then have a goal-seeking behavior (or behaviors) pick the best direction from whatever possibilities are left.

Figure 4.13 shows an example of least-commitment arbitration. A behavior that detects collision hazards activates when the robot moves too close to an object, as indicated by the dashed line. Rather than making an arbitrary choice concerning which way to move, the collision monitoring behavior computes the directions that would be unsafe. The directions marked by an X are unsafe because if the robot continued moving in any of those directions, it might collide with the obstacle. The arbiter considers motions that have not been excluded, combining such motions with the motion desired by another behavior, one that tries to move the robot toward some goal or along some path.

Summary

- Behaviors attempt to control scarce robot resources in conflicting ways.

- Behavior-based systems use an arbiter to resolve conflicts between behaviors.

- Graceful degradation enables continued robot operation in the presence of compromised sensor data.

- In fixed-priority arbitration schemes, a constant and unique numerical priority is assigned to each behavior. When behaviors conflict, the highest-priority behavior requesting control wins.

- Behavior-based systems seek to achieve robust performance using a combination of moderate-reliability systems, rather than a single very high-reliability system.

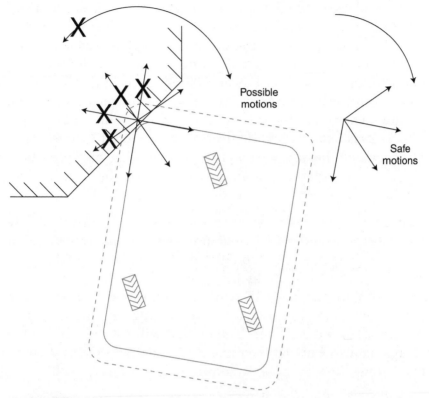

Figure 4.13

In one implementation of a least-commitment arbitration scheme, we coarsely tessellate the robot's possible motion choices. In the robot-centered coordinate system, the robot can move forward or backward, left or right, and diagonally left or right, forward or back, for eight possible translations. The robot also has two possible rotations—left or right about the center point. Hazard-detecting behaviors identify which of the possible translations and rotations are unsafe. From the remaining safe motions, the arbiter selects the translation and rotation closest to the motions desired by the highest-priority goal-seeking behavior. In the example, directions marked with an X are unsafe.

- Other types of arbitration schemes include variable priority, motor schema, and least commitment.

Exercises

Exercise 4.1 Suppose you have an apple-picking robot consisting of a mobile base and a long arm. A behavior called Pluck controls the arm. Pluck uses a camera tuned to the spectral signature of ripe apples to servo on and pick the fruit. Another behavior called Tree-nav uses a differential GPS system to navigate from tree to tree. Tree-nav outputs commands for the base, Pluck for the arm. If Tree-nav and Pluck are the only two behaviors, does your behavior-control system require an arbiter? Is the situation changed if the robot can become unstable and fall over if the arm is extended while the base is in motion? Draw a simple behavior diagram to describe the working of the system. Include any other important behaviors you feel the system would need.

Exercise 4.2   A robot has an Avoid behavior to drive it away from obstacles and a Charge-home behavior to drive the robot toward a charger. When Avoid detects the charger, Avoid tries to move the robot away. Which behavior should have the higher priority? What happens if, while heading for the charger, the robot encounters an obstacle? Is something missing from this scenario? Draw a behavior diagram that will get the robot to the charger and will not let the robot collide with obstacles encountered along the way. Specify additional behaviors if needed.

Exercise 4.3   Suppose that a robot has a cliff sensor connected to a behavior called Plunge-not and a collision detector (an instrumented bumper) connected to an Escape behavior. Which behavior should be assigned the higher priority? How should Plunge-not and Escape be written so that the robot is able to drive along a narrow path with a cliff on one side and a wall on the other side? Do you need additional sensors or behaviors?

Exercise 4.4   Outline the structure of a sensor-qualifying behavior for a cliff sensor. Implement this behavior as a finite state machine.

Exercise 4.5   Were you to insert the Charge-home behavior into the behavior diagram for the Collection task in **Figure 4.3**, where should it go?

Exercise 4.6   Devise an example in which it might be helpful to use the notion of inhibition as well as suppression.

Exercise 4.7   Invent a robotic task that might benefit from variable-priority arbitration. Outline the behavior diagram of your system. Is the benefit over a fixed-priority system significant?

Exercise 4.8   The potential field method (or a motor schema based on potentials), at first glance, might not seem to be a useful approach to solving a maze—the local-minima problem appears especially vicious for this case. However, given that you can define the potential in any way you like, how might you use a potential method to enable a robot to follow a maze? Assume that you have good global position information and that the robot can alter the potential over time.

Exercise 4.9   In retail stores, the long shiny pegs on which merchandise is often hung are usually invisible to sonar sensors. Suppose your job is to develop a robot to sweep the aisles of such a store. What combination of sensors and behaviors would you use to enable your robot to sweep as closely as possible to the pegs without running into them? (Very hard.)

5

Programming

Engineering the solution to a problem is often accomplished in one of two basic ways: either 1) appeal to first principles[1] and apply relevant theories or 2) select an appropriate example from a bag of tricks and adapt as needed. We have studied the principles of behavior-based programming, but just as a good chess player's knowledge would be incomplete without a repertoire of opening gambits, so too the robot programmer must have in his or her possession a large bag of tricks. In this chapter, we examine several common programming methods. Programmers use primitive behaviors and functions such as those described here in endless variation and countless combination to implement useful applications.

Homing Based on Differential Detectors

To home in on a destination, all the robot needs is a detector that can answer the question, "Which way should I turn?" As we saw earlier with the light-following example discussed in the section "Implementing Servo Behaviors" in Chapter 3, the differential signal from a pair of forward-pointing photocells can provide the

---

[1] First principles are the fundamental assumptions in a particular field of endeavor.

which-way-to-turn signal. Implementing such a strategy requires two sensors. Each sensor must have the property that when it is pointed directly at the home location, the signal is stronger (the output is greater) than when the sensor is pointed away from home. Ideally, the signal falls off smoothly as the angle between the sensor and homing point increases.

The differential sensors are mounted on the robot in such a way that turning the robot to maximize the signal from one sensor decreases the signal from the other. See **Figure 5.1**. Signals from the left and right sensors peak at different robot headings. The difference between the two signals is minimized when the robot points directly toward the home location. If we use this difference to determine the robot's rotation rate, then the robot does not turn when it points directly at the home location, but turns to the left

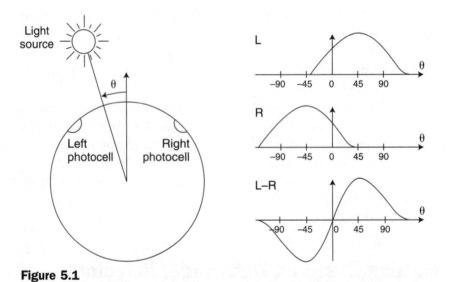

Figure 5.1

Many homing behaviors rely on differential sensors. To be most useful, sensors operated differentially should have a response that decreases monotonically as the sensor points away from the source. Here photocells are mounted at 45 and −45 degrees to the left and right of center of a circular robot. Graphs of the light intensity measured by the left, L, and right, R, photocells are shown. The difference, L–R, is used to determine which way the robot should turn to point toward the light. Positive values of L–R indicate that the light is to the left and the robot should make a positive (counterclockwise) rotation. Negative values of L–R suggest a negative (clockwise) rotation. When L–R is zero, either the robot is pointed directly at the light or the light is unseen to the rear of the robot.

when the robot points to the right of home, and turns to the right when the robot points to the left of home. This is precisely the behavior that will cause the robot to aim at the home location.

Note that any set of signals and sensors that match the requirements can be used to implement a homing system. A visible light source detected by photocells, phototransistors, or photodiodes will fill the bill. An infrared (IR) emitting beacon detected by the sort of IR receivers used in remote control systems can also be employed. And a "blob" tracker implemented with a vision system can be used in a very similar way. (The vision system must output a positive or negative value that indicates whether the blob of interest is to the left or right of center.)

A simple generic homing behavior using proportional control can be written as follows:

Behavior **Home**
    Rotation = k(L – R)
    Translation = c
End Home

It is also possible to implement homing with a bang-bang controller:

Behavior **Home-BB**
    Rotation = g *signum(L – R)
    Translation = c
End Home-BB

Here, because it offers greater clarity, we have decomposed robot motion into translation and rotational components rather than left and right wheel velocities. The methods are interchangeable (see Appendix A.) For another way to write the Home behavior, see page 56.

In the behaviors, k and g are gain parameters, L and R are the signals from the left and right detectors, respectively, and c is a positive constant. Signum(x) is a function that returns $+1$ if $x > 0$,

−1 if $x < 0$, and 0 if $x = 0$. Rotation signifies the rate at which the robot should rotate and Translation is the forward velocity of the robot. Executing either primitive behavior causes the robot to move forward as it aims toward the signal source.

It may be instructive at this point to use BSim to demonstrate the homing behavior. Make a new task with Home as a high-priority behavior and build a world with a few light sources near the center. Observe how the robot behaves as you adjust the gain and velocity parameters.

A behavior of the same form as Home can be used to realize IR *beacon following*. In this case, the home location is marked by an omnidirectional IR emitter and the robot-mounted detectors are IR proximity sensors. The trick is arranging to have the output from each detector decrease as the angle between detector and emitter increases. How can we do this?

The raw data from an IR receiver is typically a one or zero, indicating detection or non-detection of a signal. But it turns out that the detector doesn't always see the incoming signal, even when one is present. Non-detection gets worse (the signal is detected less often) as the angle between emitter and detector becomes larger. Thus, one way to obtain a differential output from your detector system is to have the beacon send a series of bursts, as shown in **Figure 5.2**. The detector software counts the number of bursts that occur within some constant standard interval. For example, when an IR receiver is pointed directly at an emitter some distance away, the receiver may detect a burst six times out of six. But when the detector is pointed 60 degrees away, it may report the signal only four times out of six. The number of times the burst signal is received by each detector during each interval then becomes the value of the L and R variables used in the Home behavior.

Line following can be implemented as a differential homing behavior. Arrange to have differential sensors detect the presence of the line to the left and right of the robot. See **Figure 5.3**. The line more nearly under the left sensor means turn left and vice versa. Translating while centering allows the line to be fol-

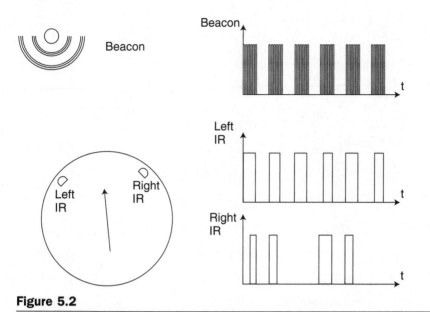

Figure 5.2

A system consisting of an IR beacon transmitting to a digital-output IR receiver mounted on a robot is able to implement differential homing. Because the left IR detector points toward the beacon more directly, the left detector senses the beacon more reliably. By modulating the beacon and counting the number of times a receiver detects the beacon during a given interval, we can determine which receiver has the stronger signal. Here the left IR detector receives the signal six times during the interval, the right IR receiver detects the signal only four times. The robot should therefore rotate left to equalize the received signals.

lowed. Because of the small capture region, line following may need additional sensors and supporting behaviors for optimum performance.

Hill climbing behaviors can also be thought of as homing behaviors where the homing signal is supplied by inclinometers. One way to implement hill climbing is to use two single-axis inclinometers mounted at plus and minus 45 degrees to the direction of travel. Turning the robot so as to minimize the difference between the two outputs will point the robot up (or down) the slope. More commonly, the inclinometers are mounted so that they measure the pitch and roll of the robot. With this arrangement, the robot climbs a hill by minimizing roll while keeping pitch positive.

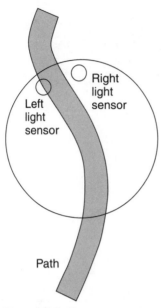

Figure 5.3

The differential signal from two downward-pointing light sensors enables a robot to follow a dark line.

Clearly, the various forms of the homing behavior do nothing more than move the robot toward a home point. Deciding when the robot is close enough to the home point that it should stop is another matter. Adding more complexity to the home behavior's body code is one way to decide when to stop. Another and often more useful way is to create a second behavior whose only purpose is to determine when the robot is close enough.

Homing Based on Absolute Position

Behavior-based systems are frequently used to accomplish tasks in which only local information is available. Because of this, you may imagine that behavior-based systems have difficulty incorporating global information. In fact, quite the opposite is true—behavior-based systems can easily incorporate absolute global information. The problem is with the cost of the information, not the ease with which it can be used. If a robot builder is willing to pay the price for an absolute positioning system, a behavior-based program can easily make good use of the data.

Several commercial positioning systems are able to provide absolute positioning information. These include systems that utilize sonar beacons or optical beacons, or systems that use laser scanners alone or in conjunction with coded targets installed in the robot's environment. The system with which most people are familiar is the global positioning system (GPS).

Let's consider an application for an outdoor robot that incorporates GPS. (GPS works well only in areas with an unobstructed view of the sky.) A GPS receiver constantly computes latitude and longitude based on timing signals received from a set of special-purpose satellites. Some receivers can be configured to output this information in any of a variety of formats. How do you make use of such information to guide your robot to a particular destination?

For simplicity, suppose that our GPS receiver outputs position information in the form of xy coordinates relative to a given origin. See **Figure 5.4**. The coordinates of the goal location (the point to which you wish the robot to move) are (X_G, Y_G) and the current coordinates of the robot (as supplied by the GPS receiver) are (X_R, Y_R). Subtracting the coordinates reveals how much the robot has to change its current position, ΔX and ΔY, to reach the goal position. Thus, $\Delta X = X_G - X_R$, and $\Delta Y = Y_G - Y_R$.

We use a coordinate system aligned with the earth's rotation as shown in the figure, with the x-axis pointing north. To arrive at the goal position, we must have the robot travel along a heading of θ relative to the x-axis. Elementary trigonometry tells us that the angle we should follow is given by the arctangent of the change in x and y position, that is: $\theta = \tan^{-1}(\Delta Y/\Delta X)$.

It is not enough to know only the absolute position of our goal and the absolute position of the robot; we must also know the robot's heading. Knowing which way to turn is the essence of homing; the required rotation is the difference between the heading that the robot is currently on and the heading along which we wish the robot to travel.

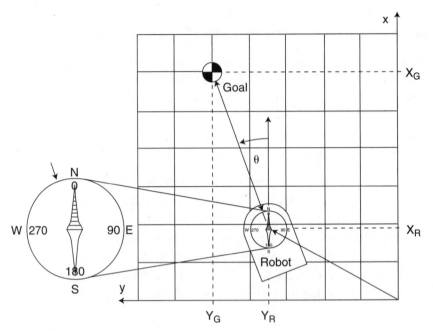

Figure 5.4

Here the robot homes on a location using information provided by a positioning system. The absolute position of the robot and the goal are used to compute a heading. An electronic compass (the compass display is enlarged at the left) enables the robot to follow the heading.

GPS provides position information, but does not directly give us the robot's heading. An electronic compass can fill in that gap. To point toward the goal, the robot spins until the heading indicated by the compass matches the desired heading. The robot proceeds by repeatedly consulting the GPS receiver for the absolute position, computing the heading from the absolute coordinates of the goal position, computing the required heading, spinning to point toward the goal, and moving forward to reduce the distance between robot and goal.

Have we now conquered the problem of moving our robot to exactly the place we want it to go? Not quite. Before we can successfully use absolute position information to home on a location, we must confront one more fiend that lies in ambush poised to bludgeon the unwary. The fiend's name is *resolution*. See **Figure 5.5**.

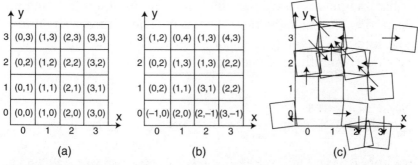

Figure 5.5

It is tempting to imagine, as shown in (a), that a positioning system establishes an orderly grid and, as our robot travels about, the positioning system tells the robot which grid cell it occupies. Unfortunately, resolution (as well as noise and other errors) limits the ability of any positioning system to function in this way. If the resolution of our positioning system is r, then at the resolution limit, there is an uncertainty of at least $\pm r$ in any measurement reported by the system. This means that, unlike the coordinates of a pixel on a computer screen, the coordinates of the robot computed by the positioning system are only probabilistic. An example of this is shown numerically in (b). When the robot occupies a particular grid cell in the real world, the positioning system may report that the robot is in a different cell. It is as if the cells have wandered from their actual positions as shown in (c)—and they wander continuously. Absolute robot positioning is built upon this uncertain foundation.

Every positioning system can accurately measure positions down to some small distance, but no smaller. For example, you can use a yardstick to measure distances as small as 1/16 inch or maybe a little better. But you cannot use a yardstick to measure the thickness of a sheet of paper. Such small distances are below the resolution limit of a yardstick. Likewise, you cannot use your car's odometer to measure the width of a basketball. Below its resolution limit, no positioning system provides meaningful information. Thus, the first question to ask of any positioning system is, "What is the resolution?"

Depending on conditions, the resolution limit of a plain vanilla GPS receiver is often no better than about 10 meters. (Although the unit may report its position down to the millimeter, the least-significant digits of the display are false precision; they are not consistent over time.) Suppose that we now attempt to use such a receiver (along with an electronic compass) to direct a robot to an absolute position. We use the following program to home on a desired *xy* location given by Dest_vec. (The nature of the undefined functions in the pseudocode below should be self-explanatory. Proper handling of vectors by the '–' operator is assumed.)

Behavior **Home_GPS**

```
    Loc_vec = get_GPS_xy()           // GPS outputs the current
                                     // location vector

    Disp_vec = Dest_vec – Loc_vec    // Displacement vector to the
                                     // destination
    Dist = magnitude(Disp_vec)       // Distance to destination
    Theta = arctan_vec(Disp_vec)     // Displacement vector gives
                                     // desired heading
    Heading = Get_compass_heading()  // Get robot's actual heading
                                     // from compass
    If (Dist ≠ 0)                    // Have we reached the
                                     // destination?
       Rotation = g1 *(heading – theta) // Compute the direction to turn
       Translation = g2 *Dist        // Compute the speed to move
    end if
end Home_GPS
```

What will happen when the robot executes this behavior? If run in simulation, Home_GPS will cause the virtual robot to turn toward the destination, move forward smoothly, and stop when the robot reaches the precise spot specified by Dest_vec. But running in a physical robot in the real world, Home_GPS will never reach the destination. Instead, the closer the robot gets to the goal, the more confused and frantic the robot will seem to become.

Far from the goal, the physical robot behaves in much the same way as its virtual cousin—moving deliberately toward the destination. But when the robot comes within about 10 or 20 meters of the goal, the resolution limit of the GPS system begins to play havoc with the rotation/translation control system expressed in Home_GPS. One second the GPS unit may tell the robot that it is in the same grid cell as the destination, the next second that it is in the cell to the left and thus should rotate 90 degrees to the right, and the following second that the robot is in the cell to the right of the goal and should spin 90 degrees to the left.

To end the robot's confusion, we must first drop our insistence that the robot reduce the position error (the distance to the goal) to zero. That is, we must establish a dead zone in the control system around Dist = 0. We will replace the statement If (Dist ≠ 0)... with If (Dist > Thresh).... Now the robot will decide that it is close enough and can stop homing when it comes within Thresh of the goal. The value of Thresh is related to the resolution limit of the GPS system. Typically, you would determine this value experimentally. Practical values may turn out to be large, possibly several times the resolution limit.

To reliably approach the goal more closely than this, you will have to purchase a positioning system with better resolution. Positioning systems with a resolution of one or two centimeters are widely available. A carrier-phase differential GPS, for example, can achieve this level of precision. Unfortunately, high-resolution absolute positioning systems that operate over a large area tend to be very expensive. This is the reason that small robots try to be clever about locating themselves. The straightforward approach simply costs too much.[2]

Avoidance with Differential Detectors

Avoidance behaviors can keep your robot safe from hazards. But to employ an avoidance behavior, you must have a sensor that can detect the hazard of interest *before* the robot is trapped or

---

[2]Can't we beat the resolution problem by averaging? The answer is a qualified yes. If the robot remains motionless in some position for a time and averages the varying positions reported by the positioning system, it will acquire a more and more accurate estimate of its true position (*if* the errors are random rather than systematic). This is how differential GPS works. One stationary receiver that has averaged to find its true position sends position corrections to a mobile GPS receiver on the robot. But the better-resolution-through-averaging solution is not free; implementing this solution requires either that the robot move slowly or that we buy a more expensive system (one that incorporates two GPS receivers, some computational hardware, and a local transmitter and receiver for passing the corrections to the robot).

damaged by the hazard. Typically programmers try to have robots avoid colliding with obstacles and falling over stairs or other drop offs. Sometimes it is also desirable that robots avoid entering or exiting particular areas specified by the user.

Perhaps the most common avoidance behavior is obstacle avoidance implemented with an IR proximity detector, IR ranging sensor, or sonar range finder. Without losing generality, we will assume a proximity detector in the behavior below. A proximity detector outputs a logical zero if no obstacles are detected within some detection distance, D_0 of the detector, and a logical one if an obstacle is present. Note, however, that the distance D_0 is a constant only for a particular combination of detector and object surface.

Suppose we have a range sensor that outputs a range value D. To make a range sensor act like a proximity detector, apply a threshold to the signal; that is, return 0 if $D > D_0$ and return 1 if $D \leq D_0$. (Unfortunately, there is no mathematical contortion that will invert this operation and convert a proximity sensor into a range sensor.) A simple avoidance behavior for a robot with left- and right-mounted proximity sensors can be implemented in the following way.

Behavior **Avoid**
    If L then
        Rotation = $-\omega$
        Translation = 0
    else if R then
        Rotation = ω
        Translation = 0
    else
        Rotation = 0
        Translation = c
    end if
end Avoid

Here L and R are the outputs of the left proximity and right proximity sensors (either 0 or 1), respectively, ω is the rotation rate of the robot, and c is a positive constant. This very simple behavior works well in environments that are populated sparsely with convex obstacles. In these cases, if the robot approaches an object on the robot's right, the robot stops, spins to the left until it no longer sees the object, then resumes moving forward. For clarity, we let Avoid command forward motion when the sensors detect nothing. More typically, Avoid would produce no output under this condition and forward motion would be commanded by a separate behavior.

Use BSim to experiment with the Avoid behavior. Make a new task using Avoid and Cruise and put a few walls into the world to give Avoid something to avoid. The gain in the BSim version of Avoid corresponds to the value of ω in the text.

If the space in which the robot operates is densely populated with obstacles or if the obstacles are non-convex, Avoid can experience a problem. The problem often occurs when the robot faces into a corner or box canyon, as it does in **Figure 5.6**(c). In this situation, the robot first detects an obstacle to the left and begins to turn right. The turn to the right causes the robot to see an obstacle to the right, so the robot turns left so that it once again sees an object on the left. The vicious cycle repeats indefinitely, trapping the robot in a corner from which it should eas-

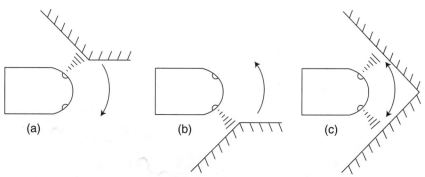

(a)   (b)   (c)

Figure 5.6

A robot executing an avoidance behavior turns away from detected objects on either side. In (a), an object detected by the left IR sensor causes the robot to turn right. In (b), an object on the right makes the robot turn left. But an inside corner, as in (c), can cause a problem for simple implementations of Avoid. Here the robot becomes trapped, spinning back and forth and never making any progress.

ily escape. This unsuccessful attempt to escape an inside corner is sometimes called a *canyon* or *canyoning* effect.

It is tempting to fix such behavior by having the robot back up whenever it sees obstacles on both sides. Doing so causes the robot to retreat from any canyon. Unfortunately, this only moves the problem back one step. Now the robot heads itself into the canyon until it sees both sides at the same time, backs up until it sees nothing, then goes forward, detects the walls again, backs up again, and so on. Once again, the robot is stuck in a repetitive cycle.

What is the fundamental problem here? Our simple servo behavior has failed us because it takes contradictory actions in similar situations. Under certain environmental arrangements, each of the behavior's possible actions brings about a situation that calls for the contradictory action. We can fix this problem either by eliminating the contradictory actions from our servo behavior or by adding state and thus incorporating a ballistic element.

Anti-Canyoning

We can construct an obstacle avoidance behavior that is not stymied by canyons if we have the robot do the same thing each time it detects an obstacle. For example:

```
Behavior Avoid_consistent
    If (L or R) then
        Rotation = ω
        Translation = 0
    else
        Rotation = 0
        Translation = c
    end if
end Avoid_consistent
```

This avoidance behavior, depicted in **Figure 5.7**, is simple, effective, and never oscillates. In Avoid_consistent the robot turns

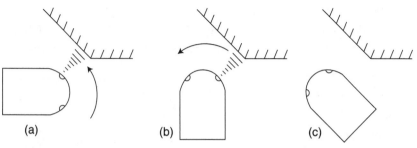

Figure 5.7

Having the robot turn in only one direction when it encounters an obstacle eliminates the oscillation that can stem from contradictory actions. But sometimes the robot makes bigger turns than we would prefer. Here the robot detects an obstacle to its left in (a). The robot begins to turn, only to sense the obstacle on its right, (b). Finally, in (c), the robot has turned far enough that it no longer senses the object.

left every time it encounters an obstacle. The robot's behavior does not generate contradictory actions and so never causes it to become stuck in an oscillation. The drawback is that if, for example, the robot encounters an obstacle obliquely on its left, then rather than making a shallow correction toward the right, it turns all the way around and goes in the other direction. The Avoid_consistent behavior is robust, but may not do exactly what you want.

The other, more dangerous, method of avoiding unhelpful cyclic behaviors is to add state—have the robot remember what it did recently so that it can forestall contradicting itself. Here is an example behavior that prevents the robot from cycling rapidly back and forth. The robot remembers the decision it made and keeps doing the same thing for a while as a hedge against rapid oscillation.

Behavior **Avoid_ballistic**
    If (test_timer(L)) then
        Rotation = $-\omega$ // Turn only one direction during interval
        Translation = 0
    else if (test_timer (R)) then
        Rotation = ω

```
            Translation = 0
        else if (L) then          // Maybe start timing an interval but only
            start_timer(L, T)     //   start a timer if none is already running
        else if (R) then
            start_timer(R, T)
        else
            Rotation = 0
            Translation = c
        end if
end Avoid_ballistic
```

As often happens when we add state, complexity begins to bloom. Avoid_ballistic presumes the existence of two functions: start_timer and test_timer. The start_timer(X, T) function creates a timer named X that runs for T seconds. If test_timer(X) is called on timer X within a time T after the timer has started, then test_timer returns true. In any other situation, test_timer returns false (including testing a timer that has not been created). If the timer is called again before time T seconds have passed, the clock starts over. That is, test_timer always returns true if called within T seconds of the most recent call to start_timer. The timer facility gives us a method for remembering and acting on events that have happened recently, e.g., the detection of an object.

The Avoid_ballistic behavior can eliminate oscillation in many circumstances. When the robot sees an obstacle on one side and turns away, only to encounter another object on the other side, it does not immediately reverse direction. Rather, the robot keeps turning in the direction it originally chose until the timer expires. By so doing, the robot may turn past the second obstacle and thus avoid the canyon.

Or the robot may not avoid the canyon. Instead, depending on the value of T, the rotation rate selected, and the arrangement of obstacles, the robot may only oscillate back and forth with larger amplitude. If we observe our robot acting in this unfortunate way, we might tweak the program, making the timer longer; i.e., increase the value of the constant T. Now the robot will turn even farther when it detects an obstacle. But it may go too far; if the robot encounters a second obstacle on the same side after it has forgotten about the first obstacle, it will just keep spinning. In that case, we will have just traded one sort of cyclic behavior for another.

There is one other trick that can sometimes help. Rather than making the time delay, T, a constant, make it a static variable, t. Let t start at some small value (when the value is near 0 the behavior is identical to the original Avoid behavior). Of course, as we saw, when the value is small, the behavior runs the risk of oscillating back and forth. But implementing t as a variable makes it possible for us to increase the value each time the obstacle detectors change their idea of which side the obstacle is on. That is, when the robot switches from seeing an obstacle on the left to seeing one on the right, we increase the value of t, but otherwise, the robot behaves as before.

The effect of this strategy is that upon encountering a canyon, the robot begins to swing back and forth, but, with each oscillation, the robot turns farther. Eventually, the robot spins just far enough to clear the canyon and is able to go on its way. We have thus added state sparingly, remembering past events just long enough to avoid an undesired repeating cycle.

The increase in the value of the timer variable must not be permanent, however. If it were, the value would just continue growing until we returned to the situation where the robot spins in place when it finds a canyon. We need to have the value of the timer variable decay back to a small number when the robot hasn't seen an obstacle for a while. (See the leaky integrator described in Appendix C.)

But even with all this complexity, the robot may remain at the mercy of the arrangement of objects in its environment. It always seems the case that with exactly the wrong combination of obstacles, the robot may go into an oscillation or a never-ending spin. As a last resort, a little randomness can save the day. If we make the turn angle random, then rather than consistently computing the wrong answer in pathological situations, eventually, by chance, the robot will do exactly the right thing!

Wall-Following with Contact Sensors

Skirting behaviors help a robot find its way amid obstacles. In a skirting behavior, a robot skirts the edge of an elongated feature. That feature might be the walls of a room, a cliff or other drop off, or even the edge of a toxic spill.

Wall following is likely the most common type of skirting behavior. See **Figure 5.8**. When a robot skirts a wall, we usually want it to move smoothly and to follow the contour of the wall precisely. The key to smooth wall following is alignment. If the robot can maintain a heading parallel to the wall as it moves for-

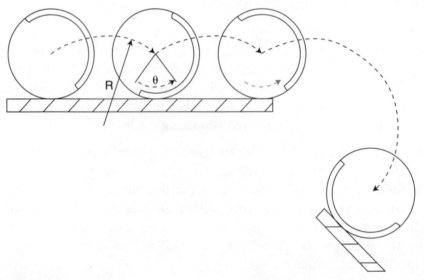

Figure 5.8

When wall following is based on contact sensing, the best the robot can do is to follow a forward arc of radius R when a wall is not detected, then spin in place through some angle θ when a wall is detected. With small values of R and θ the robot follows the wall closely but, because the robot reverses itself frequently, progress along the wall is slow. With larger values of R and θ, the robot moves forward more rapidly, but will not follow the wall as well—it may fail to pass through narrow openings in the wall as it does here.

ward, then the robot will follow the wall accurately. How can we achieve alignment?

A robot whose only sensors are bump sensors is able to follow a wall. Such a robot has no mechanism to align itself instantaneously with the wall but, by turning back and forth, it can keep its heading parallel with the wall on average. Suppose we want to follow a wall while keeping the wall to the right of the robot. A strategy that accomplishes this has the robot arc forward toward the wall until the bump sensor indicates that it is touching the wall. Next the robot spins left a constant number of degrees. The robot thus follows a scalloped path along the wall. The following behavior implements this strategy.

Behavior **Bump_follow**
```
    If (test_spin(A))           // If a spin was commanded then
        Rotation = ω            //    spin in place to the left
        Translation = 0
    else if (L or R)            // If a bump occurs on either side
        set_spin(A,θ_c)         //    command a spin
      else                      // If there was no recent bump, arc to
        Rotation = –ω           //    the right
        Translation = c
    end if
end Bump_follow
```

Here the functions set_spin(label, angle) and test_spin(label) are analogous to the functions start_timer(X,T) and test_timer(X) from the "Anti-Canyoning" section. Bump_follow works in the sense that the behavior enables the robot accurately to follow any wall or rigid object. But the robot doesn't move very quickly, the strategy doesn't make the robot look particularly intelligent, and because it repeatedly collides with the wall, either the robot or wall might be damaged (especially if the robot is big and heavy). We can do better with a sensor that lets the robot act before it actually collides with the wall.

Wall-Following with Ranging Sensors

A transverse-mounted range sensor makes it easy to follow a wall, as shown in **Figure 5.9**. In this example, our strategy is to try to maintain a constant wall/sensor distance. When the distance decreases, we have the robot turn away from the wall. When the distance increases, the robot must turn toward the wall. A somewhat subtle detail (illustrated in **Figure 5.10**) is that the measurement point must lead (or follow) the robot's center of rotation by some amount. Were the range sensor mounted on the same axis as the wheels, we would find that the measured distance offers no hint as to which way to turn to achieve alignment.

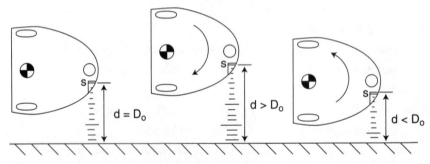

Figure 5.9

A ranging sensor, s, directed perpendicularly to the robot's direction of travel measures the distance to the wall. The desired wall/sensor separation is D_0, thus the robot arcs to the right when the measured distance is greater than D_0 and arcs to the left when the distance is less than D_0. For best results, the range sensor should lead the robot's point of rotation as shown.

Behavior **Range_follow**
    Rotation = $g * (D_0 - d)$
    Translation = c
end Range_follow

Here D_0 is the desired skirting distance, the cushion we want to maintain between the robot (or rather the sensor) and the wall. The distance that our ranging sensor measures is d, g is the gain of our proportional controller, and c is the translation velocity, a positive constant. Executing this behavior (with gain properly

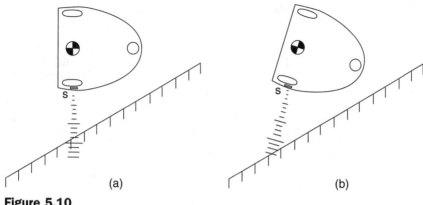

Figure 5.10

The robot in (a) is too close to the wall, as measured by the distance sensor, s, mounted on the axis of the drive wheels. But when the robot attempts to correct by turning toward the wall in (b), the measured distance *increases*. If the robot is using the Range_follow behavior, this leads to another command to turn toward the wall and eventual collision.

adjusted) gives us the behavior we want. When the distance between robot and wall is less than the desired distance (we assume the wall is to the right of the robot), the value computed for rotation is positive. The robot turns in a positive (counterclockwise) direction and moves away from the wall. When the range is larger than D_0, rotation is negative, meaning that the robot turns toward the wall.

Wall-Following with Proximity Sensors

Range sensors are more expensive and tend to be more temperamental than proximity sensors. Whenever possible, roboticists prefer to use simpler, cheaper sensors. It is common practice to build wall-following behaviors using only a single proximity sensor. The behavior that accomplishes this bears a strong resemblance to the Range_follow behavior above.

Behavior **Proximity_follow**
    Rotation = ω * (2P-1)
    Translate = c
end Proximity_follow

The result that we want is for the robot to arc left when it is too close to the wall and to arc right when too far away. The behavior Proximity_follow achieves this by setting Rotation to ω * $(2P-1)$ while translating forward at velocity c. When no wall is detected, $P = 0$ thus Rotation $= -\omega$ and the robot arcs to the left. When a wall is detected, $P = 1$, Rotation becomes $+\omega$, and the robot arcs left. Under the control of Proximity_follow, the robot acts very much as it does when commanded by the Bump_follow behavior described previously.

Carrying the analysis a little farther, we see that combining ω with c yields a radius of rotation whose center is on the left side of the robot when a wall is detected and on the right side of the robot when there is no wall. So the robot's motion consists of a series of arcs toward and away from the wall. (See **Figure 2.11**.) The smaller the latency and hysteresis in the system, the more nearly the robot's path will approximate a straight line.

You can easily implement a proximity-based wall-following task in BSim. Use the Avoid and Cruise behaviors and tune the parameters for the best compromise between rapid and accurate following.

Confinement and Cliff Behaviors

Sometimes it is necessary to confine a robot to a particular area. Behaviors that instantiate confinement are much like avoidance behaviors, but here the "object" to be avoided is the boundary separating the allowed from the restricted area. There are numerous ways to mark a boundary that the robot should not cross. Some robotic lawn mowers use a buried signal-emitting wire; the robot detects the signal and avoids the wire. It is also possible to mark a boundary with fluorescent paint. The robot carries an ultraviolet (UV) source that causes the paint to emit light in a particular range of wavelengths. A sensor also on the robot detects the characteristic glow when the robot comes near the paint, triggering a behavior that turns the robot away. Yet another way is to set up a boundary demarked by an IR beam.

A very simple strategy, shown in **Figure 5.11**, will prevent the robot from crossing a broad boundary, no matter how the boundary is marked. When the sensor detects the boundary, the robot spins in place until the sensor no longer detects the boundary. Following this strategy, the robot turns away from the boundary. Note that the sensor cannot be mounted at the center of the robot or behind the center. If the sensor is mounted at the robot's center of rotation, then the robot will spin continuously when it encounters a boundary.

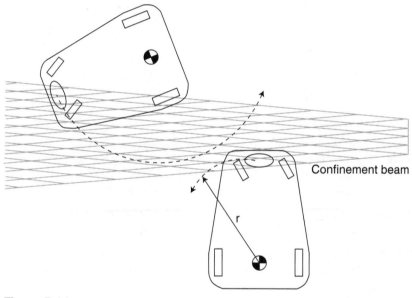

Confinement beam

r

Figure 5.11

A robot can be confined by a boundary sensor. Here a confinement sensor (the bold ellipse) is mounted ahead of the robot's center of rotation. Whenever the sensor detects the boundary (the crosshatched region), the robot spins in place—always in the same direction—until the boundary is no longer detected. Regardless of whether the robot begins on the left or right side of the boundary and regardless of the approach direction, this strategy always directs the robot away from the boundary. Note, however, that the boundary must have a detectable width greater than the point-of-rotation sensor to distance, r. A boundary region narrower than this will sometimes allow the robot to escape.

Cliff avoidance can be instantiated in very much the same way. See **Figure 5.12**. The difference is only in the number and positioning of the sensors. A robot can make do with a single confinement sensor, but may require several cliff sensors. The crucial dif-

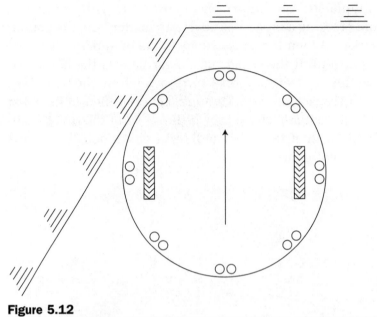

Figure 5.12

To be effective, the behavior that controls cliff avoidance must know as soon as possible when any portion of the robot overhangs a cliff. Thus a number of cliff-avoidance sensors (circle pairs) may be needed.

ference is that it is allowable for one or more of the robot's wheels to pass over the boundary as long as the bulk of the robot remains on one side. But it is not permissible for any of the wheels to go over the cliff; the robot's task will end abruptly if this occurs.

Thrashing

The canyoning effect we saw earlier is an example of a more general problem sometimes called *thrashing*. Thrashing occurs when two different behaviors are alternately given control or two parts of one behavior rapidly contradict each other. Suppose we have one behavior that drives the robot forward when a light is ahead and a higher-priority behavior that makes the robot back up when it encounters a wall blocking the path. Such a situation is shown in **Figure 5.13**. All is well until the robot finds itself approaching a light from behind a low wall. When it gets close to a wall, the robot will suddenly seem to freeze, shuddering

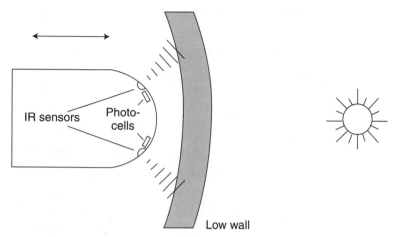

Figure 5.13

To avoid thrashing, attention must be paid to pairs of behaviors that, due to typical environmental situations, cyclically contradict each other. Here a light-seeking behavior based on the robot's photocells drives the robot forward toward the light. At the same time, a higher-priority IR sensor-based obstacle-avoidance behavior causes the robot to back away from obstacles that are directly in front of it. In the situation shown, the light-following and avoidance behaviors alternately take control, causing the robot go forward, then back up, and then go forward again, and so on.

slightly. What happens is that the light-following behavior drives the robot forward until the avoidance behavior triggers and drives the robot backward. As soon as the robot has moved backward a small distance, the avoidance behavior becomes untriggered and the light-following behavior takes over, moving the robot forward again. The arbiter alternates control between the two behaviors, and each behavior in turn contradicts the commands just issued by the other.

One strategy for overcoming this problem is to employ a cycle detection behavior. Such a behavior provides the robot with a modicum of introspection. The cycle detection behavior monitors not the external world, but rather the robot's own actions. When the cycle detection behavior senses a series of rapid back-and-forth wheel motions or just a lack of robot progress, the cycle detection behavior takes control and does something different—maybe an arbitrary spin or some random motion is commanded in order to break the cycle.

The cycle detection behavior provides a general means of escape from unexpected behavior combinations that produce undesirable results. But, when possible, a better approach is to understand the interactions between pairs of behaviors that are frequently called in close combination with each other. In our current example, we might avoid thrashing by building a separate behavior that specifically handles the case of homing and obstacle avoidance.

Another approach that can shed light on contradictory intra-behavior or inter-behavior responses is to examine a table of possible responses. If two behaviors or two conditions commonly occur together, consider explicitly what may happen. Look for response rules that rarely, if ever, lead to rapid reversals of either or both drive wheels. Consider a simple-minded rule-based wall-following behavior that uses an IR proximity sensor to detect the wall. This behavior, just like the Bump-follow behavior discussed in the section "Following with Contract Sensors," has the robot spin in place to the left when a wall is detected, but arc forward when no wall is visible. As the robot travels along the wall, the wall sensor alternately answers the question, "Is a wall visible?" Answer either yes or no. The following table outlines the robot's response.

Wall Visible?	Left wheel	Right wheel	Robot response
Yes	Backward	Forward	Spin left
No	Forward	Stop	Arc right

As the robot skirts a wall, the left wheel alternates between turning full speed forward and full speed backward. Following the wall using a contact sensor gives us no choice—when the robot strikes the wall, it has to spin (and thus periodically reverse the left wheel), but using a proximity sensor, we have a better option. Although the arc/spin strategy does enable wall following, it is not a desirable situation, as rapid reversals can stress motors and shorten their lives. A better rule would have the robot arc left then arc right. Following this strategy means that the left wheel alternates between stop and forward, rather than backward and forward. The robot now may sometimes bump the wall rather than turn away, but an escape behavior presumably already exists to handle this situation.

Escape

Generic escape behaviors were discussed in conjunction with the finite state machines in the section "FSM Example: Escape." But there is one more subtlety worth citing. Sometimes when a robot collides with a wall or other obstacle, we want the robot to rebound in a random direction. (This ability will prove useful in the next section.)

An instrumented bumper[3] can determine the approximate angle at which the robot strikes an obstacle. Suppose that the robot

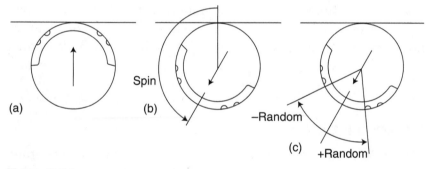

Figure 5.14

To enable finer-grained control, escape behaviors can include two extra parameters—Spin and Random. The Spin parameter causes the robot to spin a given number of degrees when the behavior is triggered. In (a), a collision triggers the behavior. The robot can be thought of as first turning in place the number of degrees specified by the Spin variable (b). A further rotation is specified by picking a random number in the range –Random to +Random (c). Thus, if the function random(Random) returns a number in the range 0 to Random, the total robot rotation is computed from: Rotation = Spin + Random – random(2 * Random).

---

[3]To determine the angle of collision, the bumper must float; that is, it must be able to move in any direction in response to an applied force. Also, there must be a means to determine the direction the bumper has moved. A pair of mechanical switches mounted on either side of the bumper is a frequently used solution to this problem. One switch closure means that the robot has collided with an object on the same side of the robot as the switch that closed. The closure of both switches means that the robot has collided with an object near the center of the bumper.

strikes the wall head on, as shown in **Figure 5.14**. If we simply choose a random number between 0 and 360 and command the robot to turn that many degrees, then, on half of its encounters with a wall, the robot will immediately collide with the wall again. Any time the robot assumes a heading between 0 and $90-\epsilon$ (where ϵ is a small number) or between $270+\epsilon$ and 360, the robot will still be aimed toward the wall, rather than away from it.

What we would like to do is to force our escape behavior to exclude the range of headings that will result in an immediate collision. Implementing Escape with two additional parameters, as shown in **Figure 5.14**, can accomplish this.

Area Coverage

Many applications demand that a mobile robot thoroughly cover a given area; i.e., the robot is required to visit, at least once, every point within some specified region. Mowing a lawn, searching for land mines, and cleaning windows are three disparate examples of robotic tasks that require good coverage.

Deterministic Coverage

Ideally, to achieve good coverage, a robot should use an accurate, high-resolution positioning system. Given such a system, a robot can cover an area deterministically. Operating deterministically, a robot always chooses the next grid cell or cells to visit based on full and accurate knowledge of all the cells visited in the past. In fact, before making its first move, the robot can formulate a complete, optimal plan of exactly the path it will take to visit every point while minimizing backtracking and revisiting already-visited cells.

Unfortunately, there are stubborn real-world problems that work against such idealized plans. As we saw in the section "Homing Based on Absolute Position," position system resolution is always an issue. Systems that exhibit high resolution are expensive—the higher the resolution, the more expensive the system—yet good resolution is typically assumed in coverage-related research work. The process a robot is to perform on a surface (mowing, say) has a fixed width, w, (the width of the robotic lawn mower's blade). Because of positioning system uncertainty, the minimum resolution we might hope to get away with is $w/2$. If our mower has a 20-inch wide blade, then our positioning system must have a resolution of 10 inches. (And that number leaves us no error margin; in practice, we would require better resolution.)

Real positioning systems have dropouts. Whether we use carrier phase differential GPS, laser scanner-based triangulation, or another method, there always seem to be areas where trees or buildings block the satellite signal or an unfortunate reflection confuses the system, causing it to supply either no information or wrong information.

Also, coverage plans that robots formulate in advance have a way of being scrambled by the unforeseen. Perhaps an obstinate dog blocks a critical part of the path the lawn-mowing robot planned to use or maybe a fenced flowerbed is not exactly where the homeowner said it was on the map the robot uses.

What happens when a navigation algorithm that assumes accurate information instead is given inaccurate information or no information? The results are disappointing—when the robot acts on wrong information, it does the wrong thing. And because the robot tends to get the wrong information every time it arrives at a particular position (where the trees block the satellite or the reflection occurs), the algorithm may systematically direct the robot to revisit areas it has already visited or, usually worse, the robot is consistently directed away from certain areas that it has never visited. The latter effect is often called *systematic neglect*.[4]

Random Coverage

In many situations, random coverage is a superior alternative to deterministic coverage. A robot that uses a random strategy to

---

[4]We might hope that some form of graceful degradation will rescue the robot in situations of systematic neglect. Indeed, as robot programmers, we should strive to achieve this. Unfortunately, devising a way to allow an absolute positioning system to degrade gracefully can be very challenging.

cover an area loses the putative efficiency of a positioning system-based strategy but benefits by avoiding the cost, complexity, and brittleness that often accompany such systems.

A random coverage strategy is just what it seems. Rather than trying to keep track of where it is, the robot simply moves randomly, changing direction whenever it meets an obstruction.[5] Just like a gas molecule captured in a container—after enough time has passed, the robot will have an equal probability of being anywhere in the area to which it is confined. See **Figure 5.15**.

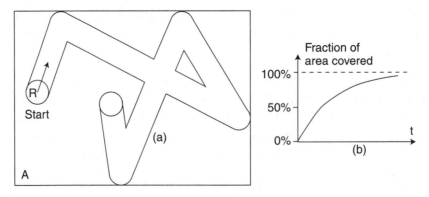

Figure 5.15

In (a), robot R beginning at Start moves randomly about the area, A, to be covered. Each time the robot encounters a wall or other obstruction, it spins a random number of degrees and rebounds in a new direction. The graph (b) illustrates how, on average, total coverage increases with time asymptotically approaching 100 percent.

Implementing a random coverage strategy is relatively simple. The coverage task can consist of as few as two behaviors; call them Cruise and Escape-random. Cruise makes the robot travel in a straight line. Escape-random can be identical to versions of Escape discussed in the section, "Escape." The BSim robot employs this coverage strategy to search for pucks while performing the Collection task.

---

[5]Changing direction only at obstacle boundaries turns out to be a more efficient strategy than a true random walk. Were the robot to change direction in open areas, it would cover the space less rapidly. This is because while the robot is spinning or arcing, it covers less area than while moving in a straight line.

The randomly bouncing robot does not know where it has been and thus cannot avoid revisiting areas already visited. This means that such a robot spends some fraction of its time in previously visited areas and the longer it travels, the larger the fraction of its time it spends in old areas, rather than new ones. Thus the rate of (new) coverage decreases as the robot operates, approximately following the formula[6]:

Coverage fraction = $(1 - e^{-t/a})$

where t is the time variable and a is the constant time it would take to cover the area deterministically without ever revisiting an already visited spot.

This equation may hold fairly well for coverage of large open areas, but often we want the robot to operate in cluttered regions. The lawn-mowing robot will encounter trees, lawn furniture, flowerbeds, and so on. Going around all the clutter slows down the mowing, but also creates a more pernicious condition—a difficulty analogous to gaseous diffusion.

As indicated in **Figure 5.16**, unfavorable arrangements of obstacles tend to confine the robot to certain areas or rough chambers. The only way the robot has to get from one chamber to another is to bounce fortuitously through an opening. If the opening between two chambers is narrow, then the probability that the robot will go through that opening is small. This means that, on average, the robot will have to bounce many times in one chamber before finding its way into the next.

Gas molecules have the same problem the robot has as they disperse to establish a uniform distribution inside a chambered container. But *given enough time*, a gas that enters such a container at one point will eventually be present in equal density at all points inside the container. Likewise, given enough time, our bouncing robot would visit each point in its coverage region an approximately equal number of times.

---

[6]For more information on search strategies, see "Randomized Search Strategies with Imperfect Sensors" by Douglas Gage. *Proceedings of SPIE Mobile Robots VIII*, Sept. 1993, Vol. 2058, pp. 270-279.

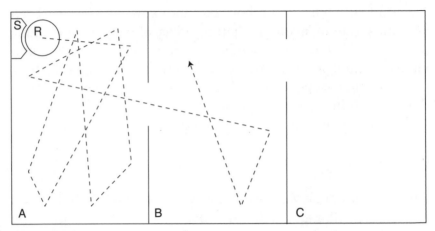

Figure 5.16

The arrangement of obstacles in a region that a robot is required to cover approximates dividing that region into a set of chambers, possibly with narrow intervening passageways. The narrower the passageway, the less frequently the robot will randomly bounce through. Yet to reach the most distant chamber, the robot must traverse all the intermediate chambers; that is, it must diffuse through the chambers, as would molecules of gas in an analogous situation. If the robot always begins in the same position (as it might if it must return to charging station, S) and if the robot can only operate for a finite length of time before exhausting its batteries, then the initial chamber A will be well covered, but the robot will rarely visit points in more distant chambers B and C.

But the robot may not have enough time. If the robot recharges its batteries at a fixed station and always begins its coverage task from that point, then it will likely provide good coverage in the chamber that contains the station. But the robot's batteries may well run down before it makes its way to more distant chambers. Thus the robot will spend too much time in the first chamber and too little time in all the others.

Researchers have devised a useful strategy to mitigate the diffusion problem.[7] Rather than bounce exclusively, have the robot sometimes follow the wall for a short distance, f. See **Figure 5.17**. Whenever the robot begins following the wall within f of an opening, it will escape one chamber and enter another. This

---

[7]See "Sweep Strategies for a Sensory-driven, Behavior-based Vacuum Cleaning Agent" by Keith Doty and Reid Harrison, *AAAI 1993 Fall Symposium Series, Instantiating Real-World Agents*, Raleigh, NC.

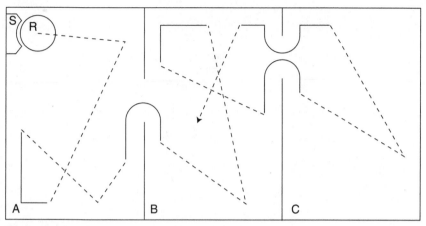

Figure 5.17

Here the robot, R, improves coverage of a cluttered area by sometimes bouncing away from walls (dashed lines) and sometimes following walls for a short distance (solid lines). This strategy has the effect of seeming to make the openings between chambers wider (if the robot begins following near an opening, it will pass to the next chamber), thereby making the robot more likely to visit chambers distant from where it starts.

simple strategy has the effect of making the openings between chambers appear larger to the robot and greatly increases the likelihood that the robot will spend an equal amount of time in all chambers.

Generalized Differential Response

We have looked at several behavior types that had differential sensor inputs and commanded differential actuator outputs. That is, we have looked at systems that possessed left- and right-pointing sensors of different sorts. Ultimately, the processed output of these sensors has been used to control the robot's left and right drive motors. See **Figure 5.18**. The values produced by the sensors have also been used as behavior triggers.

Many useful and simple behaviors can be constructed for such systems. We will now treat systems characterized by differential inputs and outputs with a bit more generality. The treatment we

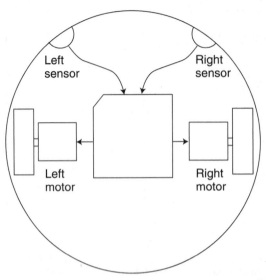

Figure 5.18

A common robotic control situation uses inputs from two sensors of the same type on opposite sides of the robot. An algorithm connects these inputs to two outputs that control drive motors on opposite sides of the mobility platform. The general linear transform provides a useful generic method for mapping sensory inputs into motor outputs.

arrive at is more complex, but also much more powerful than those we have considered before. Although there are, of course, infinitely many ways to transform inputs into outputs, there exists a particular class of transformation that is both simple and useful. Mathematical novices should steel themselves, as we are about to consider the *general linear transform*.

In many of the cases we have discussed, we have two inputs (left and right sensor values) and three outputs: rotation rate, translation rate, and trigger activation (a flag that tells the arbiter whether the behavior wants control). A concept called a linear transformation gives us a general method for transforming a number of inputs into a number of outputs. Suppose the inputs are L and R, our left and right sensor inputs, and the outputs are Translation, Rotation, and Trigger, the translation velocity, rotation velocity, and active flag (or trigger), respectively. A general method for transforming inputs into outputs is to let:

$$\text{Translation} = a_{11}L + a_{21}R + a_{31}$$
$$\text{Rotation} = a_{12}L + a_{22}R + a_{32}$$
$$\text{Trigger} = a_{13}L + a_{23}R + a_{33}$$

The subscripted coefficients (the a_{lm}s) are constant parameters.[8] First note that using this more general method, we can reproduce our earlier results for avoidance and wall following for both range sensors and proximity sensors. We do this by selecting appropriate values for the coefficients.

Recall that in our Home behavior discussion, we had Translation = c, Rotation = $k(L - R)$. We will achieve this outcome if we specify the following coefficients:

$$a_{11} = 0, \qquad a_{21} = 0 \qquad a_{31} = c$$
$$a_{12} = k, \qquad a_{22} = -k, \qquad a_{33} = 0$$
$$a_{13} = 0, \qquad a_{23} = 0, \qquad a_{33} = 1$$

The fact that a_{33} is 1 while a_{13} and a_{23} are both 0 means that the behavior is always triggered, regardless of the value of L and R. (We use the convention that a behavior is triggered whenever the value of Trigger is greater than zero.)

The actions of an Avoid behavior that triggers only when obstacles are present might be specified in a table of responses as follows:

Left sensor	Right sensor	Rotation	Translation	Triggered?	Description
0	0	undefined	undefined	No	Avoid not triggered
0	1	ω	0	Yes	Spin to the left
1	0	$-\omega$	0	Yes	Spin to the right
1	1	$-\omega$	0	Yes	Spin to the right

[8]The mathematically experienced will recognize that we have performed a matrix multiplication. We have multiplied the input vector (L, R, 1) by the 3×3 matrix **A** to yield the vector (Translation, Rotation, Trigger)^T.

This response can be implemented with the following set of coefficients:

$a_{11} = 0,$ $a_{21} = 0$ $a_{31} = 0$
$a_{12} = -2\omega,$ $a_{22} = 0,$ $a_{32} = \omega$
$a_{13} = 1,$ $a_{23} = 1,$ $a_{33} = 0$

Although we can indeed create behaviors in this way, you may wonder why we would want to go to such trouble. Note that in the case of the Home behavior, we used sensors with analog outputs. And in the case of the Avoid behavior, we had on/off sensors that just output ones and zeros. Yet we were able to implement both behaviors using exactly the same system. Just by changing the values of a few parameters, we implemented two seemingly quite different behaviors in the exactly the same way. This means that, implemented using the linear transformation formalism, Home and Avoid are not two different behaviors, they are the *same* behavior with different parameters!

So, how do we compute the matrix coefficients? Without explicitly using matrix algebra, the simplest method is first to make a table of responses as we did above, and then substitute values for the sensors to generate new equations. Finally, solve the equations for the coefficients. For example, to solve for the rotation coefficients, we could write these equations:

$\omega =$ a_{22} $+ a_{32}$ (when L = 0 and R = 1)
$-\omega =$ a_{12} $+ a_{32}$ (when L = 1 and R = 0)
$-\omega =$ $a_{12} +$ a_{22} $+ a_{32}$ (when L = 1 and R = 1)

Subtract the second equation from the third to solve for a_{22} ($a_{22} = 0$). Substitute this result into the first equation to see that $a_{32} = \omega$. Finally, substitute this result into the second equation to solve for a_{12} ($a_{12} = -2\omega$).

Avoidance and Homing Using Vectors

Earlier, in the section "Motor Schema" in Chapter 4, we looked at an arbitration method called motor schema that made use of a global vector field. Where, you may have wondered, did that global field come from? Usually, a mobile robot has no access to global information of this sort, but instead must generate its responses from local information. Robots commonly use a ring of sonar sensors to collect range data from nearby obstacles. From this data, we can construct piecemeal a repulsive vector field that can keep the robot from getting too close to any obstacle. The sorts of behaviors we have already learned about can be adapted to use sonar information; there is no need to adopt a different form of arbitration.

A simple avoidance behavior can be implemented by first taking measurements with all the sonar sensors. The behavior then makes the robot spin until the minimum range measurement is at the robot's back. Next, the robot moves forward. (To execute this strategy in the example of **Figure 5.19**, the robot would spin counterclockwise until sonar 6 moves into the position now occupied by sonar 10.) Doing this constantly moves the robot as far away as possible from all known obstacles. A variation of this approach is to take such actions only if the range reading is less than some threshold value. This prevents the robot from ever moving closer to any object than the specified minimum distance.

Another method to avoid objects using sonar sensors involves computing a vector sum. Suppose that we imagine (as described in the section "Motor Schema" in Chapter 4) that each object produces a repulsive force meant to drive the robot away. We can define such a force in this way:

$$\vec{f}_n = \frac{-\hat{r}_n}{|\vec{r}_n{}^2|}$$

where the vector $\vec{f}_n$ is the force supplied by the nth sonar sensor, the vector $\vec{r}_n$ is the distance from the nth sonar sensor to the nearest object, and $\hat{r}_n$ is a unit vector pointing outward from the

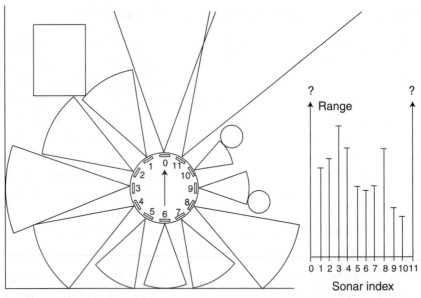

Figure 5.19

Many robots are constructed with a ring of sonar sensors that enable distance measurements in all directions from the robot. In this example, a robot with 12 sonar sensors measures the range to nearby objects. Note that sonar beams usually have a substantial width, and that the distance reported is the distance to the part of the beam that strikes a surface first (not the distance measured by the center of the beam). Sonars 0 and 11 received no echo and thus cannot compute an accurate distance. Not indicated in the drawing is the phenomenon that sonar beams striking a surface obliquely sometimes fail to return an echo. Sonar 8, for example, might report no echo.

nth sonar sensor along the center of the beam. This definition gives us a force that obeys an inverse square law. When far from objects, the force is very small, but as the robot closes in, the force rises rapidly.[9] The total repulsive force, $\tilde{f}_R$, felt by the robot is then just the sum of the forces generated by all N sonar sensors:

---

[9]There is nothing that requires us to use an inverse square law in this case. If we want the robot to react to more distant objects, we might use $1/r$, rather than $1/r^2$. Or if we want the robot to ignore obstacles until it is quite close, we could use $1/r^3$ or an even higher power or r. A force-sensing bumper is analogous to a sonar sensor that computes forces using a high power or r. The bumper completely ignores obstacles until the distance is zero, then the effective repulsive force goes to a value sufficiently large to stop the robot.

$$\vec{f}_R = \sum_{n=0}^{N} \vec{f}_n$$

To make the robot move toward a goal, we add an attractive force $\vec{f}_A$. The total imagined force, $\vec{f}_T$, is then:

$$\vec{f}_r = \vec{f}_R + \vec{f}_A$$

Unless we have an omnidirectional base, we will have to rotate the robot to point in the direction of $\vec{f}_T$ and then compute a forward velocity, perhaps just simply multiplying our imaginary force by some gain parameter.

Sometimes the goals of the robot are such that this level of complexity is needed. But often the robot is able to perform its function perfectly well with much simpler avoidance mechanisms.

Debugging

The first time you run a robot program, it will not work properly. The more complex your task, the more reliable this prediction becomes. Thus, the need to debug code is an inescapable reality of robot programming.

Fortunately, when coding errors must be found and corrected, behavior-based programs offer an important advantage. Recall that in a well-formed behavior-based program, no behavior has knowledge of the internal state of any other behavior. Also, the connections between behaviors (when they exist) are always explicit. These facts greatly assist the programmer in debugging behavior-based programs because they allow primitive behaviors to be implemented and debugged one at a time.

There is no requirement that the whole system be in place before debugging commences. Indeed, the preferred method for constructing a behavior-based robot program is to develop and test each behavior in turn. This strategy means that the physical robot is exercised early on and flaws in design or execution become apparent sooner rather than later.

Throughout the text, we have considered many example behaviors—all of them relatively simple. Correcting the individual behaviors, therefore, is sufficiently straightforward that significant debugging challenges rarely flow from this source. Rather, the hard part of debugging a behavior-based program usually involves understanding and managing the interactions between behaviors. As we saw earlier in the thrashing example, interactions between behaviors can flow through the robot and the robot's environment.

To debug behavior interactions, the programmer must identify which behaviors are interacting. It is good practice then, to build into your program some means by which the robot can indicate what it is doing. Ideally, the robot should continuously report which behavior is in control and the sensor values on which the robot is acting. BSim possesses such a facility—the space to the right of the simulated world constantly displays the state of BSim's sensors and the name of the behavior selected as winner by the arbiter. BSim also maintains a history or trace of recently in-control behaviors.

Debugging information such as this is indispensable when programming a physical robot to perform a complex task. Often an LCD screen or wireless connection to a host computer is used to display the arbitration winner and certain sensor values. For less-capable robots, tones played on a piezoelectric buzzer or patterns displayed on a few LEDs can serve the same purpose.

Summary

In this chapter, we have seen:

- To perform a homing behavior, all the robot needs to know is which way to turn.

- Differential sensors can enable homing in a simple way.

- Resolution is a key parameter that limits the performance of positioning systems.

- Avoidance can be implemented using differential detection.

- Wall following can be implemented with contact, proximity, and range sensors.

- To avoid thrashing, intra- and inter-behavior effects must be considered.

- Random coverage is best implemented using a combination of bounce and wall following.

- The structure of a behavior-based program facilitates debugging by allowing the task to be built up from behaviors one at a time.

Exercises

Exercise 5.1   Besides those mentioned in the text, what other signals and sensors can you think of that could implement homing?

Exercise 5.2   What is the simplest way to transform a homing behavior into a running-away behavior?

Exercise 5.3   In the Bump_follow behavior described earlier, we have the robot arc toward the wall, but spin in place when it touches the wall. Wouldn't the robot progress more rapidly along the wall if it were to arc rather than spin when it touches the wall? Would arcing in this situation be a good idea or a bad idea?

Exercise 5.4   Set up a task and an environment in BSim that produces thrashing behavior. (You may be able to do this using Home and Avoid or Escape plus a creative arrangement of obstacles.)

Exercise 5.5   Explain the basic problem of implementing a wall follower using a range or proximity sensor that is mounted on the same axis as the drive wheels of the robot, as in **Figure 5.10**.

Exercise 5.6   Suppose that an on-axis mounting location is the only feasible point for a wall-following sensor on some particular robot. Is there any way that the information provided by such a sensor could be used to follow a wall? If so, will this method work as well as a method enabled by mounting the sensor ahead of the drive wheel axis?

Exercise 5.7 Implement in pseudocode an Escape behavior that has the robot bounce away from a wall in a random direction, assuming that you have available a function bump_angle() that reports the direction at which the robot struck the wall. (When the robot hits the wall dead on, bump_angle() reports an angle of 0). Have your Escape behavior command turns that ensure that the robot will not immediately strike the wall again.

Exercise 5.8 In what ways does the Proximity_follow behavior perform less well than we might desire? Consider such issues as tight inside corners and sharp turns to the right. Can you think of a way to add state to this system that would improve its performance?

Exercise 5.9 Is the contact sensor-based Bump_follow behavior described early in the chapter a servo behavior or a ballistic behavior?

Exercise 5.10 Create a BSim task that exhibits the same functionality as the Bump_follow behavior of the section, "Wall-Following with Contact Sensors." Use the behaviors Escape and Cruise for this purpose. What values of parameters enable the robot to traverse the wall quickly? What values enable the robot to follow the wall accurately such that it successfully enters small passages?

Exercise 5.11 Derive the formula for the random coverage rate stated in the section, "Wall-Following with Contact Sensors." (Very hard.)

Exercise 5.12 The Home-BB behavior described in the section, "Homing Based on Differential Detectors" may have a problem when the robot is pointed almost directly at a light source. In this situation, tiny changes in the light seen by each photocell will make the robot spin one way and then the other. Add a dead band to Home-BB to eliminate this rapid cycling.

Exercise 5.13 Write in pseudocode a version of the anti-canyoning behavior that uses the leaky_integrator outlined in

Appendix C. What is the maximum value that the leaky_sum variable can attain, given the values in the text? How long will it take leaky_sum to return to zero if the robot sees no more objects?

Exercise 5.14   Using the GDR behavior in BSim, build a task that enables the robot to orbit a light source.

Exercise 5.15   Assume that you have a robot with two directional microphones mounted such that they point diagonally outward on either side of the robot. How can you use this arrangement to home on a loud sound? Suppose the loud sound occurs only intermittently. Use pseudocode to write a behavior that will get the robot to the source of the sound as quickly as possible but that will not cause the robot to wander away during long periods of silence.

Exercise 5.16   Suppose that your robot is equipped with a single sensor that can detect the concentration of natural gas present in the air. How might your robot use this sensor to home in on the source of a gas leak?

Exercise 5.17   Draw a behavior diagram and use pseudocode to write the behaviors of a robot program that will handle the situation described in the section, "Thrashing." The robot should be able to get past simple barriers and reach the light.

Exercise 5.18   Imagine a tricycle-drive robot possessing one passive steerable wheel that controls the robot's direction and two driven wheels powered by a single motor. A single light sensor is mounted on the steerable wheel and points in the same direction as the wheel. A motor rotates the steerable wheel at a constant slow rate. Given this arrangement, how can you control the robot such that the robot homes on a light source? Draw a behavior diagram and describe the behaviors needed to accomplish this.

Exercise 5.19   The BSim robot can be made to follow walls using either its bump sensor or its IR sensors. Which method performs better and why?

Exercise 5.20   Use BSim to construct a world and design your own task using at least four behaviors. Perhaps have the robot follow a light source when one can be seen and follow walls otherwise. Build the task by implementing the behaviors one at a time.

Exercise 5.21   Create a line-following task for the BSim robot. Arrange a number of light sources in an approximate circle of large radius. Select behaviors and parameter values that will enable the robot continuously to follow the circle of lights.

Exercise 5.22   Can you think of how a robot might follow a line using only one light sensor? How would performance of this robot compare to that of a robot using two or more line sensors?

Exercise 5.23   Use BSim to implement the Collection task using more than one robot. Do more robots allow the task to be accomplished more quickly?

Exercise 5.24   Build a system using BSim with several robots that try to keep away from each other and one robot that uses the Remote behavior. Use the Remote-controlled behavior to herd the other robots into a small corral.

6

Decomposition

We have studied many aspects of robot programming, but there remains a mysterious black art yet to explore. Once you have identified a task that you wish a robot to perform, how exactly, do you go about turning that task into a program? With infinitely many ways to chop a big task into smaller, more manageable pieces, where should you make the cuts? What principles and considerations should guide the decomposition of a large, complex task into a set of small, simple behaviors?

At the current state of the art, the decomposition process is not yet a science. Perhaps then, the best we can do is to illuminate key issues and considerations by analyzing the matters involved in a robotic project. The construction of a useful robot is much more than programming. Because behavior-based robots operate in the real world, they must deal with all the messy practical details of an actual, physical environment. The comfortable modularizing abstractions afforded the computer programmer are of limited help—in behavior-based robotics, everything connects to something else and everything counts. The robot's physical structure, sensor choice and placement, and program must work together to support whatever problem-solving strategy we pick.

SodaBot

To dramatize the soup-to-nuts process of creating a robot program, imagine the scenario of a struggling startup company with a cockroach problem. The new company's engineers write software, test software, and debug software day and night. Rarely leaving their cubicles, they subsist on a diet of snack food and soda. The snack wrappers end up in the trash cans, which are emptied regularly. But the soda cans are recyclable, and the recycling container is far from the places where the engineers lurk. Thus, soda cans, coated on the inside with a sticky and nutritious (for cockroaches) residue, tend to accumulate in the offices.[1] Being a disciplined roboticist, you finish your work by 5:00 PM each day and so can afford to donate a few evening hours to building and programming a cockroach-confounding robot. A well-executed robot, you reason, should rescue the harried hackers from the horde of creepy crawlies.

Stating the Problem

The crucial first step in any robotic project is to get clear in your own mind the answer to this question: What, exactly, is the problem you wish to solve? In their eagerness to plunge into the intriguing details (building and programming), novice roboticists too often give insufficient thought to this essential aspect. But without a clear and unambiguous statement of the problem, your effort at a solution will be unfocused. Time will be lost building systems that are not necessary to the actual solution while other systems that are needed go unbuilt.

In our example, cockroaches have infested company offices. So the basic problem we wish to solve, simply put, is this: We wish

---

[1]This example owes much to Jonathan Connell's "Collection Machine" project at the MIT Artificial Intelligence Laboratory. Connell actually constructed a robot that could wander about the AI Lab collecting soda cans and returning them to a depository. What Connell's robot, Herbert, lacked in mechanical and electrical robustness (the mean time between failures was about 20 minutes), it made up for in programming innovation. See *A Colony Architecture for an Artificial Creature* by Jonathan Connell, AI-TR 1151, MIT, 1989.

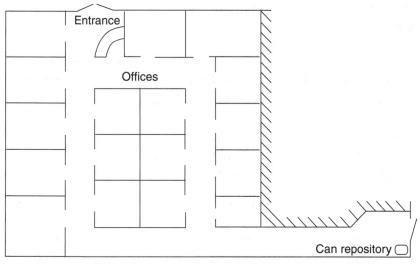

Figure 6.1

The floor plan of the company consists of a number of offices and cubicles. Cockroaches have overrun them all. The recycling repository is near the back entrance, far from the offices.

to eliminate cockroaches from the company offices. The office floor plan is shown in **Figure 6.1**.

There are many possible approaches to solving this problem. We could conceivably decide to build a cockroach-catching robot— but with millions of years of evolution working in the cockroaches' favor, that approach seems quite challenging. The genesis of the cockroach problem is the numerous soda cans, i.e., cockroach buffets, left lying around the offices. Deny the cockroaches their food source and we eliminate the problem. Thus we might consider putting a cap on each empty can to prevent access by cockroaches or we might think of injecting a bit of cockroach repellent or poison into the cans. But these solutions present their own difficulties and dangers. In this case, the obvious solution may actually be the best: program a robot to do for the engineers what they will not do for themselves. Thus, the robot's task is: Move the cans to the recycling container.[2]

---

[2]For pedagogical reasons, we have assumed here that a robotic solution is the best solution. However, when you propose a project for commercial develop-

For a human the statement, "move the cans to the recycling container" is likely sufficient to specify the task. But this declaration falls far short of the detail needed to enable robotic implementation. Humans can draw on years of experience and common sense to interpret the task description; robots have no such advantages. The problem statement implicitly relies on many human concepts—more than are immediately obvious. In order to build a robotic system capable of accomplishing the task, we must identify these unstated concepts and convert them into simple questions the robot can answer and simple actions it can take.

A human, aware of the context of the problem statement, would know that the assumed first step of the task is to search in the company offices for soda cans. But the concept of "office" holds no meaning for a robot, so we must find a way to get the robot to visit places where soda cans are likely to be.

The next stumbling block for the robot also is not acknowledged in the problem statement. Before the robot can move a soda can, it must find a soda can. And before it can find a soda can, it must be able to identify a soda can. (We'll leave aside for the moment whether the robot can reach the can it has identified.) The ability to discriminate between a soda can and a chair leg is a basic human competency so innate and universal that it is simply assumed in the problem statement. But, for a robot, deciding what is and what is not a soda can is a challenge of the first order.

The next part of the problem calls for the application of common sense, more trouble for the robot. Does moving the soda cans

---

ment, you subject yourself to the mercies of competitors, those unenlightened legions who too often have no fondness for robots. If there exist non-robotic methods that solve the stated problem at a lower cost, your robotic project is likely doomed before you begin (your business rivals and customers will see to that). In our example, we might avoid building a robot by having the cleaning crew collect the soda cans at the same time they collect the trash. The increased cost of doing this might very well be less than the cost of developing and operating the robot. Other non-robotic solutions include options like: Fine anyone caught with an empty can in his or her office. Eliminate the soda can vending machine in favor of one that dispenses the soda into cups. (The cups can be thrown in the trash.) Provide only diet soda that has no nutritional value for roaches. Building a robot is more fun than any of these, but having your product fail in the marketplace is no fun at all.

mean moving *all* the soda cans? How about cans that are completely or partly full? A human, we can be sure, would know better than to snatch away a freshly vended soda from an engineer who had set the can down for a second between sips. Perhaps the probability of premature collection is sufficiently small that we can accept a few snatched cans but, if possible, we should build into our robotic systems an analog for this modicum of common sense.

Next is the issue of moving the cans. The robot will have to pick up, push, roll, or otherwise apply force to the can in order to alter the can's position. We must choose the means. Are there constraints on how the robot should transport the cans? What if a can is partially filled with soda?

Somehow the robot must navigate itself and its payload of can or cans from wherever the soda can is discovered to the vicinity of the recycling container. Is there a particular path that the robot should or must take? What if that path is blocked?

And what about the recycling container? How will the robot identify the container? Is there only one? What if the container is not at the expected position—must the robot determine that the container is missing and avoid dropping the can on the ground? How will the robot place the can in the container?

In specifying the can collection task, we would not need to remind a human worker to avoid smashing into walls and other objects or to avoid falling down the stairs. But the need to avoid such counterproductive activities is not obvious to the robot. We must think of all the hazards the robot may face and build in systems able to deal with each specific challenge.

Also, if we gave the task of recycling the soda cans to a person, we would not expect to find that person collapsed on the floor from hunger at some later time, having forgotten to eat. But unless we make specific provision for recharging the robot's batteries, the robot will at some point stop performing its task. For truly autonomous operation, our robot will have to include a battery-recharging ability in addition to all the other things.

Now that we've taken apart the problem statement fairly thoroughly, we can see that the central competencies the robot will need to have are these:

1. Visit places where soda cans are likely to be.

2. Identify soda cans in that environment.

3. Discriminate between cans still in use and those that have been abandoned.

4. Pick up abandoned cans.

5. Navigate to the vicinity of the recycling bin.

6. Identify the recycling bin.

7. Place the can in the bin.

8. Avoid/escape hazards that might entrap the robot or cause it to fall.

9. Recharge batteries as required.

Our analysis is far from complete, but we have at least decomposed the stated task into a number of smaller more manageable pieces. The robot has some hope of being able to accomplish these subtasks. We arrived at this division by thinking carefully about the task, taking the robot's point of view, and trying to expose the hidden assumptions concerning human capabilities embedded in the problem statement. But is this an effective decomposition? We can't yet be certain, but let's proceed toward writing the program—always ready to reconsider our earlier decisions if we start to get bogged down.

Accomplishing the Task Simply

Since we believe that we have identified manageable task components for the robot, what is the simplest, most reliable way to implement those components? Our purpose is to build and program a robot on a fairly short time scale, not to engage in an open-ended research project. Thus, in picking methods to accomplish the can-gathering task, we will choose existing technology that can be adapted to our needs. To be successful, we must respect the abilities, costs, and constraints of existing technologies.

Visit Likely Spots

How can we get the robot to visit every office in search of soda cans when the concept of "office" means nothing to a robot? We might imagine that we will purchase a vision system for our robot that can identify doorways (not to mention soda cans and recycling bins). This is a more realistic option today than it once was, but if possible, let's avoid the complexity, cost, and computational requirement of such a system. If we have the robot move about randomly, sometimes following walls, sometimes crossing open spaces, then just like a lawn-mowing or floor-cleaning robot, our SodaBot will eventually investigate every office. Random search does not pretend to be an efficient method; it is not an approach that a human worker would employ, but such a method is quite appropriate for a robot. (Refer to the section "Random Coverage" in Chapter 5.) Efficiency is of relatively little concern to the robot because it has nothing else to do. Its "life" is finding and collecting soda cans. If we can eliminate the cost, complexity, and development effort required to implement a more deterministic scheme, we should do so.

Identify Soda Cans

For the robot, identifying the soda cans may be the most difficult part of the entire task. Identifying a soda can means that we must discover features that are easily and reliably detectable by inexpensive sensors. Furthermore, the combination of features we choose must be unlikely to match other items the robot will encounter in its environment. Fortunately, soda cans are quite distinctive in this regard. In the United States, at any rate, soda cans are approximate cylinders 2.6 inches in diameter and 4.9 inches high. All have a Universal Product Code (UPC) patch somewhere on the can. And all are made of aluminum.

Basing recognition on the UPC marking is appealing because it allows unambiguous identification of the can. Unfortunately, for a randomly oriented can, we should expect the code to be visible less than half the time, because the can may be faced away from the robot. Also, the code is not a point, but a patch of finite length—to identify the can, the robot would have to be able to see the entire length of the patch. Another problem is the expense, complexity, power requirements, and size of the bar code reader that the robot would have to carry around to implement the UPC solution. A method that is likely adequate for our purposes involves discrimination based on size, shape, and weight.

See **Figure 6.2**. Here an IR proximity system with one emitter and two detectors measures the approximate height of the can. When the robot is near a can, the upper receiver, A, should see nothing at the same time the lower receiver, B, detects a reflection. The logical condition, –A & B (not A and B), can be used to trigger a can-seeking behavior. If the seeking behavior fails, then probably the object was too large to fit into the gripper; thus, the object is not a can. If the object does fit the gripper, a break-beam sensor, C, indicates the presence of an object. The gripper then closes to pick up the object. If the gripper does not engage the object (as indicated by gripper sensor, D), then the diameter must be too small and the object is rejected. If the gripper does engage the object but raising the gripper makes the robot tilt forward (sensor E triggers), then the object must be too heavy (maybe we are trying to pick up a table leg with dark-colored stripes that has fooled the height detector) and the object is rejected. Only when all conditions are met does the robot conclude that it has a soda can in its grasp.

With five one-bit sensors, we have managed to identify a soda can. To the robot, a soda can is neither more nor less than the logical expression: –A & B & C & D & –E.

Abandoned versus Active Sodas

It may not actually be necessary to avoid picking up active cans. If an engineer is drinking from a can when the robot enters the office, he or she can easily shoo the robot away. But it is instructive to consider a method for distinguishing between abandoned

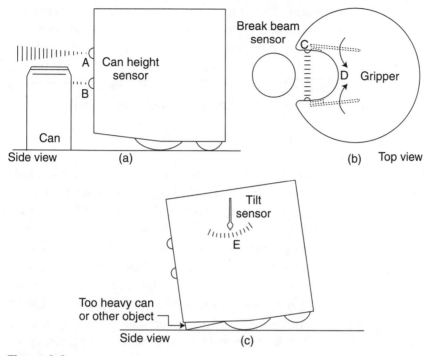

Figure 6.2

A relatively simple set of mechanisms allows us to select soda cans based on height, diameter, and weight. View (a) displays the can height sensor, composed of two detectors, A and B. A can may be present if the robot detects an object with sensor B, but not with sensor A. Top view, (b), shows the can-shaped receptacle—only an object of the proper width will fit. A break-beam sensor, C, trips when a can is present in the receptacle. The gripper (dashed) closes on the can when the break-beam triggers. A sensor, D, in the gripper triggers when the gripper closes to the proper width. (If the object in the receptacle is too slim, the gripper closes too far and does not trip the sensor.) Finally, when the gripper closes, it also lifts (c). If the object in the gripper is too heavy (a chair leg for example), then the robot will tip forward as the gripper lifts. In this case, an inclinometer, E, trips if the robot pitches forward, indicating that the object in the gripper is not suitable for transfer to the recycling container.

sodas versus ones that are being consumed. This is one of those rare cases were we are able to endow the robot with an iota of common sense.

Typically, sodas are vended cold and consumed before they reach room temperature. Thus, the only instrumentation the robot needs to determine whether a soda is still wanted versus one that has been abandoned is an inexpensive temperature sensor. With a temperature sensor installed on the its gripper, the robot measures the temperature of the can whenever the gripper closes on a can. If the can's temperature is significantly below room temperature, the robot will put the can back down. Otherwise the can will be moved to the collection point.

Can Pickup

This easiest of operations for a person can be an extreme challenge for a robot. The issue of picking up the can will force us either to engage in a difficult mechanical design and construction project or to restrict the pickup problem in a significant way. People possess dexterous, multi-degree-of-freedom manipulators with extensive force and tactile feedback. People are able to perform visual servoing on virtually any target. Whether the can is placed high on a shelf or low on the floor, fully visible or almost completely occluded, a human can easily pick up a soda can. Robots have yet to demonstrate analogous competence in real-world situations. Thus, rather than trying to duplicate the versatility of the human manipulation system, we will choose to restrict the problem.

The harried, soda-drinking engineers are unwilling to transport their empty cans to the recycling container because of the time it takes. But if it will make their offices vermin-free, perhaps the engineers would be willing to make a small concession and set the empty cans on the floor next to the wall. That easy step enables a tremendous simplification of the robotic system. By eliminating the long-reach mechanical arm that would be required to solve the general problem, we can reduce the cost and complexity of the robotic system, decrease our development time, and greatly improve system reliability. Creative simplification, when it is possible, can easily make the difference between a robot that accomplishes its task and one that is never quite ready for prime time.

Should the robot transport one can at a time or many at once? Clearly the system will be more efficient if the robot carries many cans. But in behavior-based robotics, robust but less-efficient solutions are almost always preferred over more-efficient but brittle solutions. Were the robot to carry many cans, we would first have to develop a mechanism to pick up one can. We would then have to add a cargo hold for storing multiple cans, a mechanism for moving cans into the hold, and maybe another method for getting the cans out. Also, a robot that carries many cans will necessarily be bigger than one that carries a single can. Unless something about the problem forces us to transport several cans at once, we should choose the easier-to-implement single-can solution. One simple way in which SodaBot might be designed is shown in **Figure 6.3**.

Figure 6.3

A very simple implementation of SodaBot is shown. The central cutout is just the right shape to receive a soda can.

Navigation

Navigation is a perennial challenge for robots. Outdoors, a robot that needs to know where it is can use a GPS system of one flavor or another. Indoors, GPS is of no help. Several generic indoor positioning solutions have been demonstrated,[3] but these meth-

---

[3]See "Probabilistic Algorithms and the Interactive Museum Tour-Guide Robot Minerva" by S. Thrun, et al. *International Journal of Robotics Research*, Vol. 19, No. 11, pp. 972–999.

ods tend to be expensive and computationally intensive, or they require installing targets or other equipment with line-of-sight access to the robot.

Again, we would prefer to avoid such complexity in the soda can gathering project. And indeed we are able to do so. It turns out that the recycling bin is located at the end of a long straight corridor near the back door, as can be seen in **Figure 6.1**. If we place an IR beacon close to the bin, then anytime the robot happens to enter the corridor, it can home on the beacon.[4] To get the robot to the corridor, we can either rely on its random motion (sometimes bouncing away from obstacles, sometimes following walls) or we could install a second beacon, operating on a different frequency, to mark the intersection. Either system can be made to work; which one we choose is a trade-off between efficiency and complexity.

Identify, Deposit, and Recharge

In many cases, building two or more separate, simple systems rather than a single complicated system can minimize overall system complexity. Conversely, it is sometimes possible to integrate several functions into a single device. It seems appropriate at this stage to use both techniques.

We have yet to address how the robot will identify the collection bin, how it will place the cans into the bin, or how it will recharge its batteries. Let's build a single device that can do all these things as well as support the homing beacon.

Soda cans are often collected in rigid plastic containers one or two feet high. We will assume that such is the destination for soda cans collected from the offices. The simplest robot we could devise (**Figure 6.3**) that is able to pick up and transport a soda can has no way to move that can two feet off the ground to deposit it in the bin. We could revise the robot to add this capability, but

---

[4]For cost, aesthetics, and maintenance reasons we prefer to avoid modifying office or home environments purely for the benefit of robots. Such modifications include installing beacons or visual targets. A single beacon located on the can collection structure has a low impact however, since the structure itself must be present in any case.

this would make the robot much bigger and more complex and consume more power. And why should the robot carry around its can deposition mechanism? This mechanism is useful in only one place (next to the collection bin). Why not leave it there?

We can simplify the system as a whole by leaving the robot as it is and designing a separate device to lift the cans into the bin; see **Figure 6.4**. The beacon that the robot follows to reach the vicinity of the collection bin can be attached to this device. The question, "How does the robot identify the collection bin?" is answered by the beacon. The robot must dock with the structure to deposit the can; thus, we can serendipitously have the structure provide a battery-charging function for the robot.

Avoid Hazards

What hazards will the robot encounter as it goes about its business of collecting soda cans? The most common difficulties an

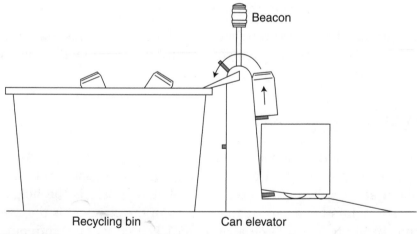

Figure 6.4

To minimize complexity, we assign some of the system functionality not to the mobile robot, but to a fixed structure. The structure provides four functions: it supports a beacon that allows the robot to home in on the structure; the structure's beacon identifies the collection point; a can elevator (implemented by a movable chain with attached paddles) provides a means for raising the cans and depositing them into the collection bin; and the structure recharges the robot's batteries. The recharging mechanism is not shown explicitly, but is embedded in the structure.

indoor robot is likely to run into are colliding with obstacles and falling down stairs. Like the BSim robot, SodaBot should have IR proximity detectors that will help it turn away from reflective obstacles before it strikes them and a full-coverage bumper to enable escape when inevitable collisions do occur. We may also need to add over-current sensing and a stasis sensor as collision sensors of last resort. If there are no stairs in the office, then cliff sensing may not be necessary.

Mechanical Platform

A standard differential drive mobility platform would seem to serve SodaBot adequately. Cleaning robots that must operate near obstacles are invariably cylindrically symmetric with the axis of the drive wheels on a diameter of the shell. This shape and configuration helps such robots escape from tight spaces because it is always possible to spin. SodaBot can perhaps afford to give objects a wider berth, but there seems no good reason for making the shape other than mostly round.

Integrating the can pickup feature with the mechanical platform dictates the central can-shaped cutout. This feature will serve to greatly simplify the control task.

Which Questions?

By the methods we have chosen so far, we have already begun to specify implicitly the questions the robot must ask as it goes about accomplishing its task. The robot will ask: Is the height sensor pointed at a can? Is there a can in the gripper? Is the can too cold to be considered abandoned? Is a collision imminent? Has a collision just occurred? Is a wall adjacent? Am I stuck? Which direction is the collection beacon? Is the battery voltage too low? Has the battery finished charging?

Every question implies an action—if this is not true, then there is no need to ask the question. Depending on the answer to a question, the robot will do one thing or it will do a different thing (or possibly nothing). If the robot decides that the can height sensor is pointed at a can, it will immediately attempt to maneuver itself to grasp the can. If the answer to the voltage-too-low question is yes, the robot will head toward the charger.

Every question the robot asks it asks for a specific purpose. A common mistake made by beginning robot designers is choosing the sensors first and the solution second. Effective design requires the opposite order. First decide how the robot will solve the problem, then the questions the robot must ask, and finally the sensors the robot will need to answer the questions. But be prepared to iterate if the sensors you would like to use are not available or are too expensive.

Which Sensors?

Figure 6.5 presents a pictorial diagram of the major systems that the robot will need to perform the can collection task in the manner just described.

To sense hazards, the robot requires an instrumented bumper. The drive motor stall sensors provide protective backup sensing in case the bumper sensor generates a false negative. IR obstacle sensors are provided to enable the robot to sense obstacles before it collides with them and to follow walls. The IR-based can height sensor lets the robot know when a soda can may be near. The break-beam sensor that crosses the can receptacle informs the robot when a can is positioned properly for gripping. The grip sensor along with the tilt sensor helps the robot qualify that the object it is gripping is a soda can.

The battery voltage sensor alerts the robot to the need to return to the charging station. The homing sensor (perhaps composed of two or more symmetrically arranged IR detectors) enables the robot to find its way to the can collector/charger. And the docking sensor lets the robot know when it has reached the correct position to release a soda can and to recharge batteries.

It may turn out that more sensors are required. Maybe the robot will need shaft encoders on the drive motors so it can drive

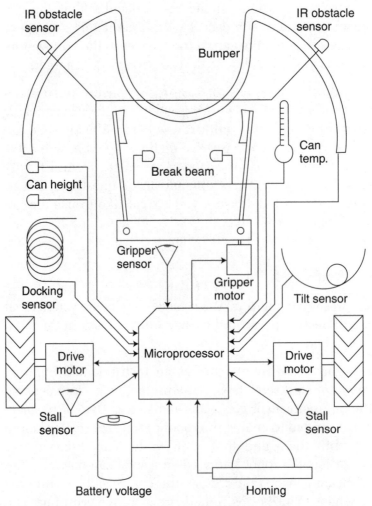

Figure 6.5

The major systems of SodaBot are shown. Drawing a graphic such as this often aids in the design process. The drawing reminds the designer of all the sensors and actuators, their approximate relationship, and all the input/output lines that will be required.

straight—or maybe not. When they are required, additional sensors or actuators can be added to the pictorial system diagram.[5]

---

[5]Many good books describe the construction of sensors and circuitry for robots. Consult titles in McGraw-Hill's Robot DNA series as well as: *Build Your Own Robot* by K. Lunt, *Sensors for Mobile Robot* by H. R. Everett, and *Mobile Robots: Inspiration to Implementation* by J. Jones, A. Flynn, and B. Seiger.

Building Behaviors

The actions the robot must take are fairly well defined. We have chosen a physical structure and a set of sensors. It is time to devise the behaviors that will bring together all the pieces of the robot and cause it to act in the way that we desire. The robot's overall behavior will emerge from the primitive component behaviors. So, what component behaviors will we need?

The scenario we have imagined (the overall behavior that we want to emerge) goes like this: The robot wanders about, sometimes traveling in a straight line across open areas, but spending much of its time following walls. When the robot happens upon a soda can, it takes some action to force the can into the gripper cutout. Ideally, the robot is able to detect the can from some distance—the farther away it can detect a can, the more efficient the robot's search. When a non-cold can is secured in the gripper, the robot begins to wander about, perhaps now looking for long straight ways. If a can is present in the gripper and the robot detects the beacon on the collection structure, then the robot begins to home on the beacon. When the robot notices that it has docked with the structure, it stops moving and releases the can from the gripper. When the robot notices that the gripper is empty and that battery charging is not required, the robot begins to search for another can. Whenever the battery voltage is low, the robot begins to search for the structure. After docking with structure, the robot remains motionless until charging is complete. If collisions occur, the robot must escape.

SodaBot Behaviors

An approximate behavior diagram that should accomplish the can collection task is shown in **Figure 6.6**. Despite the complexity of SodaBot's task, fixed-priority arbitration can be used.

Starting at the lowest priority is the Cruise behavior. Other behaviors will become active when their trigger conditions are met, that is, when they encounter the situation for which they were designed. But until that happens, Cruise keeps SodaBot

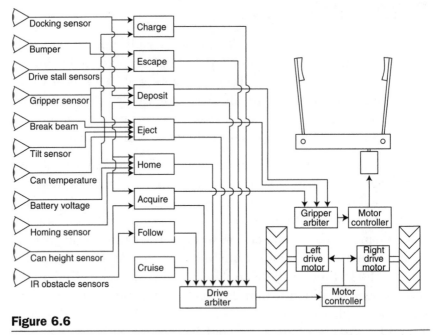

Figure 6.6

This behavior diagram indicates all the major components needed to make SodaBot collect and deposit soda cans.

moving forward, thus providing the other behaviors with fresh opportunities.

After cruising in a long straight line for a while, the robot will encounter a wall or other obstacle. Follow activates at that time. The length of time that the robot follows a wall after Follow becomes active may be a random quantity or it may prove necessary to make the wall-following time depend on the robot's situation: e.g., is there a can in the gripper? But it is while the robot follows a wall that it is most likely to encounter an empty soda can—triggering the can height sensor.

When the can height sensor triggers, the Acquire behavior becomes active; see **Figure 6.7**. If the robot is following a wall and encounters a soda can, the can height sensor triggers at a fairly well-defined point, point (b) in the figure. The opportunistically discovered relationship between can and robot allows the robot to acquire the can either by following a simple turning pattern or possibly by using the can height sensor to

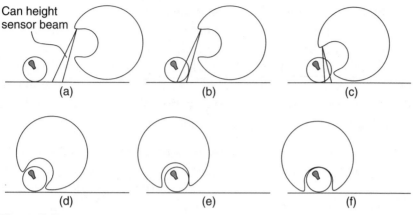

Figure 6.7

The Acquire behavior directs the robot to capture a can, as indicated in the sequence of drawings showing the robot from above. In (a), the robot follows a wall to its left. The can height sensor triggers in (b) when the lower beam detects an object, but the upper beam does not. Acquire then takes control, causing the robot to arc forward in such a way that it tends to capture the can, as shown in (c), (d), and (e). Finally, the robot drives forward (f) and activates the gripper to hold the can firmly and lift it off the floor.

servo on the can. When the can is enveloped by the receptacle, thus tripping the break beam sensor, Acquire commands the gripper to close on the can.

With the can in the gripper, the Home behavior will become active if the beacon is within sight. If not, the robot will wander around until the beacon does become visible. The Home behavior will also trigger if battery voltage is low, regardless of whether a can is present in the gripper.

There is always a possibility that the robot will attempt to collect something other than a soda can. The Eject behavior exists to prevent the robot from trying to carry a chair leg or a cold (still in use) can to the collection point. To accomplish this, Eject looks at the gripper sensor, tilt sensor, and can temperature sensors as well as the break-beam sensor—for redundancy, we might also have Eject look at the drive motor stall sensors. If any of the sensors has a value that indicates the robot is not carrying an abandoned soda can, Eject commands the gripper to release the can, the robot to back away, then turn to face a new direction.

Eventually the robot arrives at the collection structure carrying a can. The docking sensor trips when the robot has homed to the correct position. At that time, the Deposit behavior has the robot open its gripper, releasing the can to be lifted by the can elevator. After the break-beam sensor has cleared (as it does when can elevator lifts the can for deposit into the collection bin), Deposit can release control of the robot, allowing the robot to go look for another can.

If the robot has arrived at the collection structure with insufficiently charged batteries, then the Charge behavior forces the robot to remain motionless until the charger has restored battery voltage to the proper value.

Is that all? Not quite. We've taken our gedankenexperiment about as far as it can go. Were we now to attempt actually to build SodaBot, we would inevitably run across some surprises. The real world is rich in detail and unexpected interactions. Pure thought may be sufficient for success in mathematics but it always seems to fall flat in robotics. The decomposition we have chosen, the sensors we specified, and the behaviors we have engineered are just the starting point. As we try to construct SodaBot and fill in the details of its program, reality will intrude.

There are different arbiters for the drive motor and the gripper—it is possible that commands sent by one behavior will win at one arbiter and commands sent by a different behavior will win at the other. This might cause the robot to behave badly in some circumstance or it might not. If this is a problem, additional structure, perhaps a winner line, may be needed (refer to **Figure 4.5**). Maybe the beacon homing sensor will not allow precise enough alignment with the collection structure and we will have to add a short-range final approach sensor. Maybe the Acquire behavior is overburdened; perhaps a second behavior that controls just seating the can in the can receptacle and managing the gripper must be designed. Possibly having the robot just wander around until it can see the homing beacon makes the robot so inefficient that its collection activities cannot keep up with the engineers' soda drinking habits. It might then be necessary to add an intermediate beacon or two to get the robot back to the collection bin faster.

Despite these and other problems that are likely to develop, the framework is versatile and malleable and should endure. (See the exercises at the end of the chapter.)

Robot Recap

Our analysis has been a painstaking one. When designing robotic systems, we are often forced to concentrate on matters a human would consider too trivial to bother with. But there is no other choice if we are to discover the simplest way to build and program a robot. And without a steadfast pursuit of simplicity during the design phase, you will later find yourself explaining what your robot was supposed to do, rather than watching it run.

In seeking simplicity, we were not tempted to incorporate embellishments—features that might add cachet, but do not directly contribute to the solution of the problem at hand. SodaBot has no wireless connection to the Web, no speech recognition system, not even an LCD display. It just collects soda cans. SodaBot respects W. Grey Walter's admonition: "Creatures with superfluous organs do not survive; the true measure of reality is a minimum."[6]

The system we settled on is not a general one. We did not design a universal trash collection robot, or a robot transporter, or even a pick-up-small-things robot. We focused on a specific application and exploited as many features of that application as we could to generate a particular solution—one that does not require any breakthroughs in computer vision, or manipulator performance, or machine learning. Thus we have hope of actually being able to build and benefit from the can-collecting robot today; we need not await the advancement of robotic science.

The methods SodaBot employs are local methods rather than global ones—SodaBot doesn't know where the soda cans are, doesn't know where the can collection point is, and doesn't even know where *it* is. Global methods hold the promise of improved

---

[6]See W. Grey Walter, *The Living Brain*, W. W. Norton & Company, Inc., 1953, p. 131.

efficiency and greater generality, but such methods tend to be brittle. In many if not most robotic applications, global information that is wrong is worse that no global information at all. SodaBot relies only on local sensing, measurements that it can make itself and can make continuously—a much more robust strategy.

Principles

Additional principles that SodaBot attempts to follow include the following.

First Do No Harm

Programmers of behavior-based robots search ceaselessly for certain sorts of methods. These are methods that, when they function correctly, help the robot do its job, and when they fail, do no harm. Graceful degradation exemplifies this approach. When a sensor returns accurate data, the robot does its job more effectively, and when the sensor falters, other sensors and behaviors act to prevent failure of the whole system.

Tilt the Playing Field

Always seek behaviors that nudge the system in the right direction. The real-world behavior of insects offers us a good example of this principle. Watch a line of ants carrying food back to their nest. They do not behave deterministically. Sometimes they go the wrong way, sometimes they drop the food and go searching for more; if two or more ants are cooperating to carry food, they sometimes push or pull in opposite directions. But on the whole, despite their frequent missteps, the ants accomplish their task. And that is key; if the system makes forward progress, even just 51 percent of the time, then eventually the task will be accomplished.

Prefer Robust Chaos to Brittle Determinism

Efficiency is often purchased at the price of brittleness. Relying on random search, SodaBot makes the opposite trade-off.

Summary

Robots are deficient in most of the abilities that humans take for granted. Robots do, however, possess a few talents that humans lack. Because of their particular weaknesses and strengths, robots cannot typically perform tasks in the way a human would. If we wish to decompose a task in such a way that it can be accomplished by a robot, it is necessary first to develop a deep understanding of the task. Implicit reliance on unique human abilities assumed in the task statement must be exposed. Methods to accomplish the essential components of the task must then be found. These methods should play to the robot's strength and need not resemble methods a human would employ.

Choosing a good decomposition for a task means:

- Decide first *exactly* which problem you will solve.

- Analyze the problem statement, searching for hidden assumptions concerning human competencies that robots may not possess.

- State simply the minimum set of competencies the robot will need to achieve its task.

- Look for methods that will enable each competency using existing technologies.

- Develop a clear understanding of the questions the system must ask to accomplish its task.

- Match sensors to questions.

- Write behaviors that implement the methods chosen and connect the behaviors to fixed priority arbiters.

- Assume sensors will generate errors. Look for a structure that promotes graceful degradation.

- Accept methods that, on average, advance the task.

- Strive for robustness first, and efficiency second.

Exercises

Exercise 6.1   The collection structure is immobile, but its software control can nevertheless be implemented as a behavior-based system. Describe how the collection structure functions. Draw a behavior diagram and specify the operation of each behavior.

Exercise 6.2   Assume the office building where SodaBot operates includes one or more sets of stairs. Add a cliff-avoidance behavior to SodaBot's behavior diagram.

Exercise 6.3   Suppose we want to prevent SodaBot from entering certain areas. Identify a sensor system compatible with the office environment that can perform this function. Add the area-exclusion behavior to the behavior diagram and write the behavior in pseudocode.

Exercise 6.4   Suppose that after building SodaBot, it turns out that the homing beacon is not sufficiently precise to enable the robot to dock with the collection structure. Specify a sensory system that will make docking possible and add it into the behavior diagram.

Exercise 6.5   Imagine that the preliminary version of SodaBot has trouble finding its way from the offices to the collection structure. Outline how a multi-beacon system would work. The robot should find its way from one beacon to the next so that it can reach the collection structure more reliably.

Exercise 6.6   Suppose that we also want SodaBot to collect juice bottles. The bottles are the same diameter as the soda cans but taller. How should we modify the system to collect bottles?

Exercise 6.7   The juice bottles from the previous exercise should be deposited into a different repository. What additional modifications will you make to allow the soda cans to be placed in one collection bin and the juice bottles in another? Specify new sensors, behaviors, and any auxiliary equipment.

Exercise 6.8   Think of some method (using a new sensor if necessary) that will allow SodaBot to know whether it is in an office or a corridor. How, exactly, can the robot use this additional information?

Exercise 6.9   Assume that you have available the capability described in the previous exercise. Specify a behavior or behaviors that will improve the robot's can-collecting performance.

Exercise 6.10   Books in a library are effectively lost if they are in the wrong place on a shelf. Design a robot that traverses the stacks of a library and identifies misplaced books. Assume that each book has an identifying bar code on its spine.

Exercise 6.11   Use BSim to instantiate an inverse Collection task. Have the robot push pucks from a central area to the walls.

Exercise 6.12   Add a second robot to the system in the previous exercise. But have the new robot attempt to collect pucks while the first disperses them. Which robot is more effective at accomplishing its task?

Exercise 6.13   Does SodaBot work efficiently enough to accomplish its task? Assume that there are E engineers each drinking s sodas a day and that SodaBot takes on average time f to find a soda, time c to return to the can collection structure, and that SodaBot must spend one-third of its time recharging its batteries. Develop expressions showing how many soda cans are generated per day and how many SodaBot can expect to collect. (Do you need additional assumptions?) Based on the office drawing (**Figure 6.1**), make reasonable estimates for the variables and determine if SodaBot can keep up with demand. If it cannot, what is the simplest way to improve the situation?

Exercise 6.14   Choose a task that might be accomplished by a simple robot and repeat the steps outlined in this chapter to design an autonomous robotic system that could accomplish that task. Try to choose a task that does not require sensors or technical abilities that do not yet exist.

7
Physical Interfaces

Because the various facets of a robot are intertwined with each other, a book about robot programming cannot be about *only* robot programming. As we have seen, robotic systems are intrinsically cross-coupled and interconnected, with each part depending on and affecting every other part. Given these facts, it is imperative that you, the robot programmer, develop a clear understanding of whatever sensor you choose for your robot. You must appreciate how sensors work and especially how they fail. Developing a robot program often takes the form of an iterative search, in which the programmer seeks the proper balance between the needs of the algorithm on one hand and the abilities of available sensors on the other. The algorithm you choose to employ will dictate the sensors you select; but often the limitations of existing sensors will force you to modify your preferred algorithm or replace it altogether.

The details of sensors cannot all be idealized or abstracted away. To make effective sensor choices and to use the chosen sensors well, you must be cognizant of several things. First, what does the sensor actually measure? The robot needs to ask particular questions; the sensors you specify will answer those questions. But a casual or cursory assessment of a sensor's abilities leads to disappointing results. For example, robots commonly use sonar-ranging sensors as virtual yardsticks to measure the distance between the robot and an obstacle. However, from a physical standpoint, a sonar sensor does not measure distance, it measures time—the interval between the emission of an ultrasonic ping and the moment an echo is received. Under the proper circumstances, this time can be accurately interpreted as the distance between the robot and the object at which the sensor is pointed. In non-optimal circumstances, the "distance" reported by the sensor can be completely unrelated to the true distance.

Answering the question of what a sensor actually senses leads directly to the questions of under what conditions it is appropriate to use the particular sensor. In what ways is the sensor likely to give misleading results? Can we expect false positives, false negatives, or both? Can the robot determine when the sensor should not be trusted? And most important, is there a way to ensure that robot behavior degrades gracefully when the sensor supplies misleading data? Each type of sensor technology, each different use to which you put a sensor will generate different answers to these questions.

The size and weight of a sensor and the computational load that sensor imposes on the robot's processor are also important. These considerations constrain your choice of possible sensors.

Finally, the practical roboticist must ask how much the sensor costs. A sensor that produces exactly the information your robot needs is of no more than academic interest if purchasing the sensor would exhaust the funds needed for securing all the robot's other components.

This chapter provides only an overview of the complex relationship between sensors and programming. We discuss a number of sensor modalities, but do not attempt exhaustive coverage of all types of sensors. For a more complete presentation from a sensor-centric perspective, see *Sensors for Mobile Robots* by H. R. Everett, AK Peters, Ltd., 1995. Sensors have multiple uses; they are categorized below in terms of common use in robots.

Collision Sensors

The most pertinent force with which a mobile robot must concern itself is the force between robot and environment. The robot/environment force is non-zero anytime a collision occurs between the robot and an object in the environment. The robot itself generates the force of collision through the torque supplied by the drive wheels.

If wheel/floor traction is high and the object with which the robot collides is heavy, the wheels will stall during the collision; that is, the wheels stop turning even though the robot continues to apply power. **Figure 7.1** illustrates the possibilities. If traction is poor, the collision with a heavy object will stop the robot, but the robot's wheels may not stall; instead, either or both wheels may continue to spin, slipping on the floor. If the object is light in weight, the robot may impel the object—both object and robot will move, but the robot will perhaps move more slowly or consume more power than before the collision.

Dealing with these three possible outcomes of a simple collision necessarily complicates the operation of your program. Your

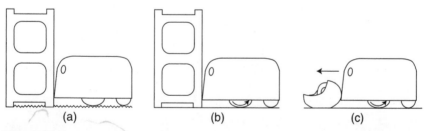

(a)   (b)   (c)

Figure 7.1

Collision with an object can produce at least three distinct outcomes. In (a), the object is immovable and the robot's drive wheels have good traction. The bumper compresses, the robot stops moving, and the drive wheels stall. In (b), traction is sufficiently low that colliding with the object stops the robot's forward motion, but the wheels continue to spin. In situation (c), the object is light enough that the robot can push it along. A properly functioning bumper will always detect situation (a). In situations (b) and (c) however, the sensitivity of the bumper determines whether the bump sensors accurately report a collision or generate a false negative.

program may need to provide for the possibility that the robot's actual situation is more complex than is reported by any single collision-detecting sensor. What will or should the robot do when this happens?

Bumpers

An instrumented bumper provides the most direct method for detecting force between robot and environment. The simplicity and straightforwardness of the bump sensor provide one of robotics' rare joys. A properly designed bumper gives a satisfyingly unambiguous answer to the robot's question, "Have I collided with an object?" Of all the sensors on which a programmer might depend, the bumper is typically the most reliable.

Although random failures can occur, false negatives and false positives generally result from poor design rather than from any inherent characteristic of bump sensors. The most common bumper design shortcoming is incomplete coverage. **Figure 7.2** shows two ways in which coverage may be incomplete. Even though in (a) the bumper occupies the forward-most points of the robot, colliding with an overhanging obstacle can generate a false negative. Also, a robot that has a forward-mounted bumper

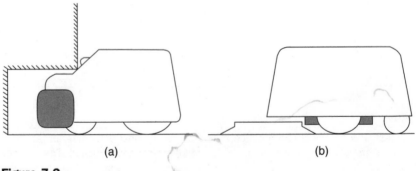

(a)   (b)

Figure 7.2

A bumper can sense only what it touches. In (a), the bumper (indicated by shading) instruments the forward-most points of the robot, however, the bumper does not fully cover the robot. Thus the collision with the overhanging portion of a cabinet goes undetected. In (b), the robot's undercarriage is poorly designed. A low threshold passes under the robot's bumper and snags on a portion of the drive system (indicated by shading) causing another undetected collision.

but no rear bumper runs the risk of a false negative bump indication anytime the robot backs up.

Obstacles that pass under the bumper can also generate false negatives. For practical purposes, there must be some clearance between the bottom of the bumper and the floor. This fact constrains the design of the robot's underside. The robot's undercarriage and mobility system must be engineered in such a way that an object small enough to pass under the bumper cannot snare the robot; see **Figure 7.2**(b). A low object catching the underside of the robot is effectively a false negative indication from the bumper—the robot has collided with a low object, but the collision sensor (the bumper) is silent on the matter.

There is one source of false negatives that is beyond the reach of clever bumper design. If the floor is slippery, the robot may not be able to generate sufficient traction to activate the bump sensor. This fact provides another example of cross-coupling between robotic systems. The failure of the bumper to activate may mean that your robot needs tires with better traction, not that you should redesign the bumper.

Sometimes other systems must be called on to mitigate the effects of false negative reports from the bumper. A drive-wheel stall detector can indicate that the robot is stopped. This provides collision detection information, although typically only one bit of information is available. That is, a bumper can be designed to report if a collision has occurred on the left, right, or center (even greater resolution is possible with more complex instrumentation). A stall sensor, on the other hand, tells only that the robot is not moving, providing no information about the relative location of the collision-causing obstacle.

Stasis sensing can also sometimes be used as a supplement to bump sensing. (For an earlier example using a stasis sensor, see the section, "Graceful Degradation," in Chapter 4.) A stasis sensor gives an indication of whether the robot is moving by looking for some type of change. Often a stasis sensor is a virtual sensor. For example, a cleaning robot intended to cover a small room will regularly bump the walls or other obstacles in the room in an effort to clean as close as possible to all obstruction. Under these circumstances we can expect the bump sensor to activate on a regular basis. If the robot's wheels have been turning (not stalled) for a long period of time but the robot has detected no collisions, it might be the case that the robot has collided undetectably with some object. If it decides that this is the case, the stasis sensor can trigger an escape maneuver. Thus, paradoxically, the absence of a collision can indicate a collision.

Poor bumper design can also make possible false positive indications. Because of its inertia, a spring-mounted bumper moves relative to the rest of the robot chassis any time the robot starts or stops suddenly. If the bumper's relative motion is large enough, a false positive collision indication will occur. This is more likely to happen if a heavy bumper is coupled to the robot chassis with weak springs and if the robot is capable of rapid acceleration.

Relatively weak springs are usually desirable because the weaker the spring, the more sensitive the bumper is to small forces. This is the fundamental trade-off a roboticist must make in designing a bumper. Strong springs make it possible for the robot to avoid false collision detections due to bumper inertia, at the cost of not being able to detect small forces.

The programmer can lessen the likelihood of false positive indications due to bumper inertia by limiting the robot's acceleration and deceleration. If the false positives tend to occur after a rapid stop (when the bumper oscillates briefly), the program can have the robot remain halted for a time, ignoring collision indications, until bumper oscillations damp out.

However, treating the symptoms of poor bumper design with clever software is not the best approach. If false negatives and/or false positives become a problem, the careful roboticist will redesign the bumper and other systems as needed so that erroneous indications do not occur in the first place.

The cost of an instrumented bumper (compared to many other sensors) is usually quite low. The simplest implementation consists of an electrical switch attached to a moving bumper in such a way that the switch is activated whenever the bumper is displaced from its neutral position. In terms of information gained for price paid, a mechanical collision sensor is one of the most cost-effective sensors you can buy.

Stall Sensors

Small mobile robots most commonly use permanent magnet direct current (PMDC) motors to power their drive wheels. PMDC motors require electric current in proportion to the torque they produce. Torque, and thus current, is at a maximum when voltage is applied, but the motor is not turning. Maximum current flows for an instant each time the motor begins to turn from a dead stop; maximum current flows continuously if an external force prevents the motor from turning. Sample graphs are shown in **Figure 7.3**. A robot's drive motors can be halted if the robot collides with an obstacle and the wheels have good traction. (If the wheels have poor traction, the robot will stop, but the wheels will spin, slipping against the floor.) Thus, maximum current flowing for a relatively long time suggests that the robot has suffered a collision.

Compared to an instrumented bumper, stall sensing is a blunt and not always reliable method. But it can be a good choice for a collision sensor of penultimate resort. Motor stall sensors can produce both false negative and false positive indications of collision. **Figure 7.3** gives some indication of the reason for this. Ideally, a stall sensor would simply monitor the current through the motor and declare a stall condition if the current exceeded some preset threshold. Unfortunately, this is not possible.

Each time the robot begins to move again after stopping, maximum current briefly flows through the motors as they come up to speed—instantaneously indicating a stall condition. Thus, before the stall sensor software declares a collision, motor current must be high (above threshold) for some length of time. The proper time interval requires a trade off. A long delay means that when a collision actually does occur, the robot will be allowed to push against an object with motors stalled. A short time delay means that the robot will often incorrectly interpret startups and temporary high-load conditions as collisions.

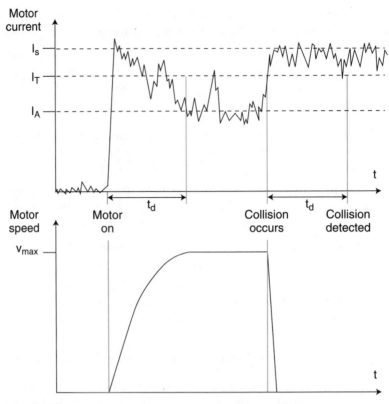

Figure 7.3

In the upper and lower graphs, motor current and motor speed, respectively, are plotted against time. When the motor begins to turn, at time Motor on, current spikes to a maximum value, I_s, the stall current. Current then decreases as the motor speeds up to its commanded value. After a time, t_d, the motor reaches the desired speed and motor current attains a steady state average of I_A. At time Collision occurs, the robot bumps into an object, forcing the wheels to stop spinning. Current increases to I_s and remains at that level. The program can infer a collision if motor current exceeds a threshold value of I_T for a time greater than t_d. But the program must always allow the time t_d to pass before it can decide that a collision has occurred. As indicated, noise can greatly complicate the interpretation of the current data.

The exact current threshold to choose is also problematic. When a robot climbs a slope or travels over certain surfaces, e.g., shag carpet or sand, the current needed to maintain velocity can be quite high—close to the stall current. Set the threshold too low, and the stall sensor will trigger when the robot is operating normally, too high, and stall will never trigger. The noise produced by all inexpensive PMDC motors is also a complicating factor, as transient noise spikes can be interpreted as collisions.

Poor robot traction produces false negative stall indications. When the robot is wedged against an object with its drive wheels spinning, current does not reach a maximum, so stall does not trigger.

Collision sensing by means of over-current detection is a much more reliable proposition if you use highly efficient drive motors. Unfortunately, such motors are expensive. Barring better motors, the best approach is to tune the stall sensor conservatively to avoid false positives and to rely on another system (e.g., a stasis sensor) to deal with occasional false negatives.

Like bump sensors, stall sensors also tend to be quite low in cost. Often such sensors can be implemented with a few inexpensive electrical components.

Stasis Sensing

As described above, a stasis sensor is a sensor that detects whether the robot is moving. There are many ways to create stasis sensors, both virtual and physical. One simple way to engineer a virtual stasis sensor is to build a behavior that monitors one or more physical sensors to see whether their values change over time. A bumper, an IR proximity sensor, a cliff detector, a photocell, and many other types of sensors change their value if the robot is in motion, but constantly output the same value if the robot stops moving. Drive wheel shaft encoders are not useful for stasis detection because the wheels can spin in a low traction environment even if the robot is snagged and motionless.

Virtual stasis sensors are free. You've already paid for the physical sensors—stasis sensing can be had for the effort it takes to write a behavior.

Avoidance Sensors

Relying solely on collision sensors to keep your robot safe is not the best strategy. Often a better approach for the robot is to attempt to avoid hazards while they are still some distance away. Various sorts of light-based and acoustic-based sensors are commonly used for this purpose.

Infrared Proximity Sensors

Near-infrared sensors (typically utilizing wavelengths in the 880 to 980 nm range) are both common and inexpensive. Nearly every remote control system used in the home has one. Robotic proximity detection systems based on this technology consist of an IR emitter and a detector. The emitter and detector are pointed outward from the robot. If an object is present in the intersecting fields of view of the emitter and detector, radiation reflected from the object will strike the detector. The absence of reflected radiation is interpreted to mean the absence of obstacles. See **Figure 7.4**.

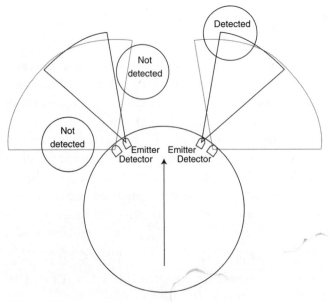

Figure 7.4

The infrared proximity sensor is one of the most common avoidance sensors employed by mobile robots. Such sensors can be used to detect obstacles and other hazards. The emitters/detector pairs that implement proximity detection each have finite fields of view. Obstacle detection occurs only when an obstacle occupies the intersecting portions of the fields of view of the emitter and detector. Therefore, individual IR proximity sensors invariably have blind spots.

To improve the performance of such systems given the presence of stray IR from other sources, the output of the emitter is typically modulated with a carrier wave of from approximately 38 kHz to 56 kHz (noise sources are unlikely to be modulated in this way). IR receivers with built-in circuitry for detecting a single frequency in this range are common. For our purposes, we need concern ourselves only with the output of the detector.

Most IR receivers output a low signal when an object is within detection range and a high signal when no object is present. Some sorts of objects fool IR detectors. These include objects that are too shiny, too absorptive, or that present too small a cross section to reflect sufficient IR from the emitter back to the detector. Examples are illustrated in **Figure 7.5**. Relying on an IR proximity sensor as the only object detection system inevitably leads to disappointment.

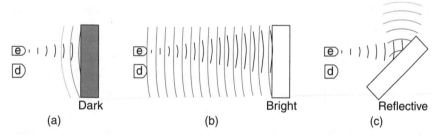

(a)   Dark   (b)   Bright   Reflective   (c)

Figure 7.5

An IR proximity detector can be thought of as answering the question, "Is an object nearby?" The dividing line between what is considered nearby or far away depends strongly on the color and texture of the object surface. In (a), IR radiation from the emitter, e, strikes a dark, highly absorptive object. No signal returns to detector, d, and the sensor reports that no obstacle is present. In (b), the bright reflective object *is* detected despite the fact that it is more distant than the dark object in (a) that was not detected. In (c), a nearby reflective object is not detected because the object directs most of the radiation away from the detector.

Multiple IR emitters and detectors may reduce the number of blind spots. Using many elements may be feasible because emitters and detectors are quite low in cost. But regardless of the effort put into the implementation, the system is always at the mercy of the environment. Dark, shiny, or small objects will more often than not cause the detector to generate false negatives. Sunlight or other bright light shining into the detector can saturate the internal components, also leading to false negatives.

IR proximity detectors less commonly generate false positive indications. False positives that do occur in a properly working system are generally due to IR noise from unexpected sources. Fluorescent lamps are sometimes the culprit here. In principle, the detector cannot know when its output is unreliable. However, positive indications that do not persist are likely to be noise glitches.[1]

Infrared Range Sensors

An IR range sensor can be had for a cost only a few times higher than an IR proximity sensor. Both digital and analog outputs are available. Such sensors have the advantage that they can return a distance, not just an object-is-present or object-is-absent indication.

Because they are based on triangulation rather than reflected intensity, IR range sensors are less sensitive to object color than are IR proximity sensors. **Figure 7.6** diagrams the popular GP2D02 unit made by Sharp Electronics Corporation. Just like IR proximity sensors, however, IR ranging sensors can produce erroneous results. Dark, shiny, or too-small objects may return no signal, and objects positioned near the limits of sensor range (both far and near) can produce inaccurate and inconsistent readings. In addition, compared to proximity sensors, range sensors can require a significant time to operate. A fast-moving robot that relies only on IR range data may not get the information it needs in time to avoid a collision under some circumstances.

---

[1] Ed Nisley has created an informative exposition of IR sensors showing how they work, how they fail, and how you can make your sensors work better. It is called *IR Sensing for the Bewildered* and can be found in zipped form via a link at: http://www.trincoll.edu/events/robot/. See also "Above The Ground Plane: IR Sensing" by Ed Nisley, *Circuit Cellar Magazine*, August 2003, pp. 40–43.

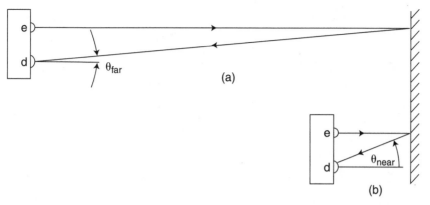

Figure 7.6

An IR range sensor determines the distance between sensor and object using triangulation. The emitter, e, projects a spot of IR light on the object whose range is to be determined. A lens in the detector, d, casts an image of the emitter-created spot on internal photosensitive components of the detector. These components are sensitive to the position (and hence the angle) of the spot's image. The angle is interpreted as a range. In (a), the small angle θ_{far}, implies a long distance while the large angle θ_{near}, in (b), corresponds to a short distance. The output of the sensor may not be linear over the entire range and the values reported may be unreliable when objects are too near or too far away. Like the IR proximity sensor, the IR range sensor may fail to detect objects of certain textures and colors.

Sonar Sensors

Sonar ranging sensors are effective at sensing large objects oriented perpendicular to the beam at a distance that is not too large and not too small. Sonar rangers can give unexpected results in other situations, as shown in **Figure 7.7**. In the figure, the sonar returns a distance that is unexpectedly short in (a), fails to detect an obstacle altogether in (b), and returns a distance that is too large in (c).

Sonar rangers can provide useful information, but allowances must be made for the ways in which such sensors can fail. Usually this means that the robot must employ non-sonar sensors as backups.

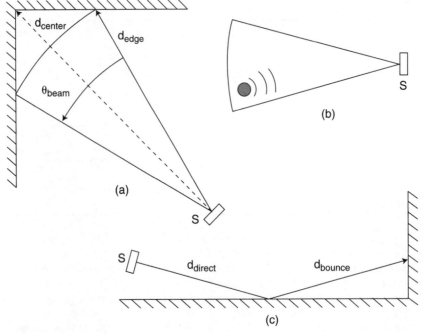

Figure 7.7

Sonar time-of-flight sensors are used for determining longer ranges than can be easily measured with common IR sensors. But acoustic sensors have many well-known foibles. The finite width of the beam, θ_{beam}, produced by sonar transducer, S, in (a) can cause inaccuracies. Here the center of the sonar beam is directed at a corner a distance d_{center} from the sensor. But the sensor reports the smaller distance, d_{edge}, because the measured echo is generated by the first surface any part of the beam encounters. Often, as in (b), the echo returned by small objects has insufficient amplitude to be detected. Because the sonar beam strikes a smooth wall at a shallow angle in (c), the beam bounces forward rather than back. But a strong echo is returned from the second wall encountered. Thus the distance reported by the sensor is $d_{direct} + d_{bounce}$ rather than the actual distance to the obstacle d_{direct}.

Range-Sensor Considerations

Suppose your robot design calls for range sensors. What are the maximum and minimum distances that such sensors should be able to measure? The answer depends critically on the purpose for which the sensor data is collected. If the range sensor is intended only to enable collision avoidance, then the maximum distance the sensor must be able to measure is the robot's stopping distance plus a safety margin. The stopping distance is the distance that the robot will move after it has decided that it needs to stop. This is a matter of physics and sensor latency.

Suppose your robot travels at a maximum velocity of v_m and can achieve a maximum constant deceleration of a. From Newton's equations of motion, we know that $v_m = at$, or $t = v_m/a$. Here t is the time the robot will need to stop. The distance, s, the robot travels while trying to stop is $s = 1/2\ at^2$. Even if your interest is robot programming rather than robot design, it is still a good idea to plug in some typical numbers to get a feel for what these equations mean.

Let's assume that the robot moves at about walking speed, say 3 feet per second, and that by applying the brakes, the robot can stop with a one g acceleration, one times the force of gravity. Then the time it takes the robot to stop is:

$$t = 3\ [\text{ft/s}] / 32\ [\text{ft/s}^2] = 0.09375\ [\text{s}]$$

The robot can stop in about a tenth of a second. The distance the robot travels in this time is:

$$s = 1/2\ {}^*32\ [\text{ft/s}^2]\ {}^*0.12 = 0.16\ [\text{ft}]$$

Thus the robot stops in about 0.16 feet or just under two inches. If the robot can only manage 0.5 g of deceleration,[2] then the stopping distance works out to less than seven inches. If the purpose of the ranging sensor is only to enable the robot to avoid collisions, then the robot can make do with a very short-range system—it has essentially no need to know about objects that are, say, 10 feet away. Unfortunately, it has been common practice in robotics to gather huge amounts of information and then throw away all but a tiny portion—meaning that designers purchase information they never use.

---

[2]This is perhaps a more reasonable number since, for a small robot, the maximum deceleration force often depends mostly on the coefficient of friction between the robot's tires and the surface on which the robot travels.

Homing Sensors

A homing sensor provides the robot with a means to reach some destination. Typically, the destination is marked in a unique way—perhaps by a beacon, light, or color.

Photocells, Phototransistors, and Photodiodes

One of the simplest ways to mark a special location is to attach a visible light. The simplest sorts of light sensors measure the intensity of light striking a single element sensor. In this class are photo-resistive elements (also known as photocells), photo-transistors, and photodiodes. The output of straightforward circuits containing such sensors is an analog voltage whose value is a function of light intensity. The light intensity reaching the element is related to the intensity of the light source, the distance to the source, and the angle of incidence between the source and detector.

From physics, we know that light intensity at a detector positioned far from the source obeys an inverse square law. That is:

$$I \propto 1/r^2$$

where I is the measured intensity of the light at the sensor and r is the distance between light source and light sensor. This is diagrammed in **Figure 7.8**. Also, the measured intensity can be a function of the angle between sensor and light source. Depending on the details of the sensor and associated lenses, we may be able to assume Lambert's cosine law, that is:

$$I \propto \cos\theta$$

Does this mean we can assume that the voltage, V, output by our sensor is just:

$$V = V_0\cos\theta/r^2$$

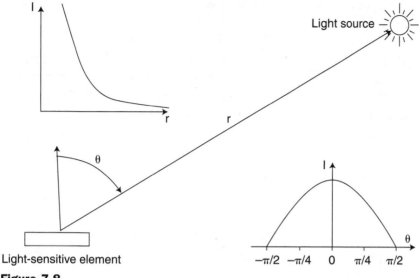

Light source

r

r

θ

I

θ

−π/2   −π/4   0   π/4   π/2

Light-sensitive element

Figure 7.8

The output of a light sensor is related to the intensity, I, of the light source, the source-to-sensor distance, r, and the angle, θ, between source and sensor. These relations enable us to write programs allowing a robot to home on a light source or avoid light sources.

where V_o is the voltage in an initial configuration? Unfortunately, we can make no such assumption. The relationship between the intensity of light falling on the sensor and the voltage output by that sensor and associated circuitry can be a complex one. (The voltage depends on the linearity of the sensor and the circuitry.) The best result we are likely to achieve is that the voltage increases monotonically as light intensity goes up. Usually this is enough for purposes of homing on or avoiding a light source.

Simple light sensors are among the most reliable of sensors. It is the interpretation of the data they deliver that can be problematic. Is the robot homing on the lamp above the recharging station or is it rather homing on the shaft of late afternoon sunlight streaming through the window?

Are all photocells the same? Indeed not; different photocells produce different signals when struck with the same light. Does this matter? Usually it matters less than you might think; see **Figure 7.9**. The mismatch between a pair of photocells used to implement homing means that when the robot thinks it is pointed directly at the light, it is actually pointed to one side. Does this imply that the robot will never arrive at the light? No, because when the robot moves forward, the imbalance increases and the robot turns to correct. Even when the mismatch is large, the effect is not to make the robot miss the light; rather, the robot spirals toward the light instead of approaching directly.

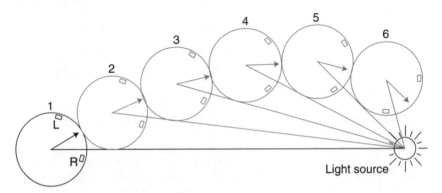

Figure 7.9

Even fairly crude, mismatched photo sensors are able to achieve homing. Here a robot attempts to home on a light source. Unfortunately, the sensitivity of the left and right photocells are very different—so much so that when the robot attempts to aim at the light, as it does in 1, it actually points 30 degrees to the left of the source. The robot thus moves to 2, where it again orients itself to point 30 degrees away from the source. The robot proceeds in this way from 3 through 6. Even though the robot is never able to point directly toward the light, it is nevertheless able to spiral in until it comes arbitrarily close.

Coded Beacons

Compared to visible light, a less ambiguous choice for marking a home location is a beacon. Using photocells to home on a lamp leaves the robot completely at the mercy of the environment. Visible light sources are very common, and if your robot happens to get a glimpse of one, it will go astray. A coded beacon, however, can be made unique.

Coded beacons (like the one SodaBot used in Chapter 6) typically consist of an omnidirectional IR source modulated in some particular way; see **Figure 7.10**. One or more receivers on the

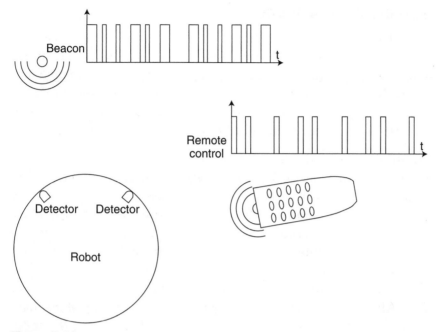

Figure 7.10

A homing beacon can be designed to emit a series of IR pulses; the robot responds only to these pulses. The particular pulse train chosen for the beacon is unlikely to match the emissions of any other device in the robot's environment, a remote control for example.

robot look only for the modulation pattern of the transmitter. This method is one of the most reliable homing means available because there is little chance that stray IR from some other source will consistently fool the system.

The ways that coded beacons can fail include:

- The area is saturated by noisy IR emissions from other sources, such that the robot cannot identify the beacon amid the noise.

- Obstacles in the environment prevent the robot from seeing the beacon much of the time.

- The sun or another source of very bright light saturates the detector onboard the robot, making it impossible for the detector to sense any signals.

195

Pyroelectric Sensors

Pyroelectric sensors are transducers that convert radiant heat into an electric signal. Typically, such sensors are sensitive to the longer wavelength far-IR radiation emitted by warm-blooded animals. Pyroelectric sensors are often used to enable a robot to identify that a person is nearby or to home on or follow a person. Such sensors are commonly found in motion-sensing burglar alarms.

Affordable pyroelectric sensors (those that are not cooled to cryogenic temperatures) detect differences in radiant heat, rather than absolute levels of heat. Thus people- or animal-detecting pyroelectric sensors can fail if the background temperature matches the temperature of whatever the robot is designed to detect. For example, an agricultural robot that relies on a pyroelectric sensor to avoid collisions with livestock might fail to detect an unfortunately positioned cow in the heat of a summer day. Also, pyroelectric sensors generally respond most strongly to the hottest source in their field of view. This makes the robot much more likely to home on the fireplace than on the person warming beside the fire.

Color Blob Sensors

Recent advances in inexpensive electronic cameras[3] make such devices suitable for use as homing sensors on robots. One of the best-developed vision algorithms is the color blob tracker. To use such a system for homing, it is necessary to choose a particular color and train the camera to that color. The camera can then be made continuously to output the coordinates of the centroid of the matching color in the camera's visual field.

Vision is an active area of research and improvements can be expected. But the caveats to use of such system include:

- Cameras have a finite frame rate (how quickly the camera captures and analyzes snapshots). If the tracked color moves too quickly, the robot won't be able to follow.

---

[3]See, for example, the inexpensive CMUCam vision system: http://www-2.cs.cmu.edu/~cmucam/.

- The camera's ability to identify color is much inferior to human abilities. Changes in illumination (moving from an area lighted with incandescent lights to one lighted by fluorescent lamps, for example) can cause the camera to lose the tracked object.

- The color of the object on which the robot homes must be unique. Other objects with similar colors will appear just as appealing to the robot as the object the programmer intends the robot to track.

Magnetic Sensors

Magnetic detectors (Hall-effect sensors, for example) can be both inexpensive and sensitive. However, for homing purposes, such sensors are typically limited to very short ranges, typically a few inches or feet. Range is limited because magnetic sources produce a dipole field whose intensity falls off as the inverse cube of distance—much faster than intensity from a point source of light decreases. Thus, such sensors are more commonly used for docking, rather than homing.

Dead Reckoning and Navigation Sensors

Dead reckoning and other navigation sensors are used to drive the robot to some location without an explicit marker on the place where the robot is going.

Shaft Encoders

The most common sensor used as a position aid on a robot is almost certainly the drive wheel shaft encoder, as shown in **Figure 7.11**. Shaft encoders measure the rotation of a motor or wheel and do so quite reliably. It is the interpretation of the shaft encoder data that can sometimes lead the robot astray.

The signal from the shaft encoder tells the robot in which direction, how fast, and how far the associated drive wheel has turned. Ideally, robot motion can be inferred from wheel motion,

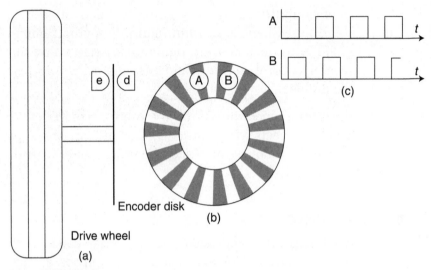

Figure 7.11

Common shaft encoders consist of a striped disk attached to the shaft of a motor or wheel. An IR emitter/detector pair looks through the striped disk. A side view of this arrangement is shown in (a); the emitter and detector are labeled e and d, respectively. As the shaft turns, the disk spins, and the stripes alternately allow or block transmission between the IR pair. A circuit monitors the pulses thus generated, each pulse corresponding to a small movement of the wheel. To determine the direction of wheel rotation, two IR emitter/detector pairs are needed. The pairs are installed such that their signals are 90 degrees out of phase. This is shown in (b), where the two channels are labeled A and B. Graph (c) shows the output as the wheel turns. Wheel direction is determined by whether signal B leads or lags A by 90 degrees.

but in practice, this relationship is uncertain. In the best case, the robot's position, as computed from shaft encoder information, contains an unknown amount of error. The error component never decreases, rather the uncertainty in our knowledge of the robot's true position just becomes greater and greater the farther the robot moves. In the worst case, when wheels slip, there is no relationship between wheel motion and robot motion. Thus, the robot cannot reliably know its position using shaft encoder information alone. (For additional information about shaft encoders, see the section, "Limits of Dead Reckoning," in Appendix A.)

Inertial Sensors

In theory, it is possible to capture the signal from inertial sensors—accelerometers and gyroscopes—integrate the signal over time and thus compute the position of the robot. In practice, the output of inexpensive inertial sensors drifts to such an extent that the robot's position cannot be reliably determined in this way.

However, the robot's attitude in relation to the gravity vector can be reliably measured with low-cost accelerometers. Also, inexpensive gyroscope chips can faithfully report rapid changes in the robot's rotation rate. These abilities allow the construction of balancing robots, among other things.[4] If you attempt to construct such a robot, keep in mind a fundamental result from physics: it is impossible for an inertial sensor to differentiate between forces caused by gravity and forces caused by changes in robot velocity due to the action of the robot's motors. That is, inertial and gravitational acceleration are indistinguishable.

Compasses

Relatively inexpensive electronic compasses are very effective at determining the local vector toward the north magnetic pole. However, such measurements are easily disrupted by the operation of the robot's motors. Attaching the compass to a mast that extends above the robot can mitigate this problem. Most static magnetic fields caused by metallic robotic components can be accounted for through careful calibration.

Unfortunately, the presence of nearby ferrous material in the robot's environment cannot be compensated for in a simple way. Such material always threatens to affect the robot's notion of north. This condition makes electronic compasses unreliable when used inside most buildings.

However, the complementary data supplied by a compass and by a robot's shaft encoders provides us with an example for demonstrating how two systems can be combined to generate information with greater reliability than either could manage alone.

---

[4]See for example: http://www.dragonflyhollow.org/matt/robots/firemarshalbill/.

Magnetic compasses tend to be globally accurate. That is, at most points within a large area, the compass will report an accurate heading. But there are always isolated spots near ferrous objects when the compass can report incorrect results; errors of as much as 180 degrees can easily occur. Orientation based on information supplied by shaft encoders tends not to exhibit sudden large errors; instead, error creeps in slowly as the robot moves—and heading uncertainty never decreases.

Suppose we wish to arrive at the best possible estimate for θ_{rec}, the heading we will use for dead-reckoning computations. The two sources of heading information we have are θ_{enc}, the heading supplied by the encoders, and θ_{comp}, the magnetic heading measured by an electronic compass. The quantity θ_{enc} is computed by subtracting the right encoder reading from the left and multiplying by a constant. Thus θ_{enc} tracks the rotation of the robot (when the robot turns left by say, 30 degrees, θ_{enc} will increase by 30 degrees), but θ_{enc} has no particular relationship to north. If we want θ_{rec} to represent the robot's true heading and if θ_{rec} is derived from θ_{enc}, we will have to add a correction:

$$\theta_{rec} = \theta_{enc} + \theta_{corr}$$

What is the value of the correction term, θ_{corr}? It is just the difference between the true heading of the robot and the heading given by the encoders alone. Thus, if we make a single measurement at a point far from any interfering ferrous objects, we can write:

$$\theta_{corr} = \theta_{comp} - \theta_{enc}$$

where θ_{comp} is the heading reported by the robot's electronic compass.

By itself, the equation $\theta_{rec} = \theta_{enc} + \theta_{corr}$ leads to unsatisfactory results because θ_{enc} keeps drifting. The farther the robot has traveled since computing θ_{corr}, the less accurate the heading will be.

So how about if we periodically update θ_{corr}? That too is a problem because we might happen to measure our correction term just as the robot passes a ferrous object—then θ_{comp} will be wrong and the heading θ_{rec} will be meaningless.

But because θ_{comp} is usually correct, what we *can* do is to take a running average as the robot moves (see the section, "Running Average" in Appendix C). Thus our best estimate for robot heading becomes:

$$\theta_{rec} = \theta_{enc} + \overline{\theta_{corr}}$$

Note that $\overline{\theta_{corr}}$ represents an average over distance rather than a time. That is, we take a new sample $(\theta_{comp} - \theta_{enc})$ and feed it into the running average computation whenever the robot has moved some distance, d, since the last sample was taken. This gives us the best heading we can wring from our sensors and situation. The value θ_{enc} drifts slowly as the robot moves, but is corrected by regular readings of the electronic compass. The electronic compass gives an occasional inaccurate reading when the robot comes within a few feet of a ferrous object, but by averaging over a long distance, the effects are damped out.

GPS

For information regarding global positioning system receivers, please review the material in the section, "Homing Based on Absolute Position," in Chapter 5.

Summary

Every sensor has its own particular problems. Even so, the litany of woes presented in this chapter, regarding the use of sensors, is not meant to be discouraging. Rather, these cautions should increase your resolve thoroughly to understand the sensors that you incorporate into a robot and to plan for graceful degradation.

Exercises

Exercise 7.1 Describe how one might use polarizing material to make a visible light-emitting beacon unique.

Exercise 7.2 The text suggests that shaft encoders attached to drive motors cannot be used directly as stasis sensors. But can you think of a way that such encoders could provide an indication that the robot is not moving as commanded? (Hint: How does the robot react immediately after power stops flowing to the drive motors?)

Exercise 7.3 Consider a stasis sensor based on bumper activity. What factors determine how long the robot should wait after its most recent collision before it decides that it may be stuck?

Exercise 7.4 Suppose that the mass of your robot's bumper is m, the spring that holds the bumper away from the robot has spring constant k, and the bumper switch trips if the bumper deflects a distance s or greater. What is the maximum acceleration that the robot can have if inertially induced false positive collision indications are to be avoided?

Exercise 7.5 Why does electrical noise generated by the drive motors complicate the computation of collisions based on motor current? See **Figure 7.3**. What steps can you take in software to mitigate the noise problem? Does your solution have any unwanted side effects?

Exercise 7.6 Using many IR emitters and detectors mounted on the front of your robot, you can eliminate most (but not all) blind spots. Where, in relation to the robot, are the blind spots that cannot be eliminated in any practical way? What sorts of objects are most likely to penetrate undetected into the robot's blind spots?

Exercise 7.7 Suppose it takes 100 mS to get a new reading from an IR range sensor. If your robot can brake with an acceleration of 0.6 g, what is the maximum speed at which your robot should be allowed to move?

8

Implementation

Abstract notions about how to design software are all well and good, but at some point, you have to write actual code. This chapter presents one comprehensive example of how to implement a behavior-based robotic system. The system is a basic one, but it incorporates all the software elements needed by any behavior-based robot. These components include: a scheduler, a coherent method for constructing behaviors and specifying their priority, a scheme for connecting behavior outputs, and an arbiter. We begin by considering the robot's abilities and structure then show how its behavior-based system is implemented.

RoCK Specifications: The Goal of a New Machine

RoCK, an acronym for Robot Conversion Kit,[1] is a self-contained electronics and sensor package designed to convert any of a wide variety of RC cars into a robot. See **Figures 8.1** and **8.2**. The purpose of RoCK is to provide users with an inexpensive path to

---

[1]Ben Wirz and I submitted RoCK as our entry into *Circuit Cellar* Magazine's Design Logic 2001 contest. (We received a runner-up award.) For a more complete story of RoCK, including the electrical design, please refer to the series of four articles we wrote for the April through July 2002 issues of *Circuit Cellar*.

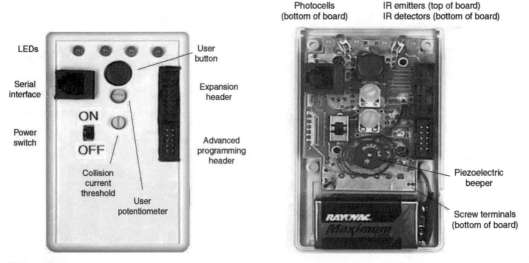

Figure 8.1

The RoCK electronics package is built into a standard plastic enclosure. RoCK's external appearance is shown on the left; on the right, the top has been removed, revealing the internal electronics board. Four LEDs in a row across the top illuminate to indicate RoCK's status. The user potentiometer allows the user to select which task RoCK will perform and to alter the value of behavior parameters.

building a robot and to demonstrate the power and versatility of behavior-based robotics.[2] RoCK comes complete with a number of built-in, user-selectable tasks. Many of the behaviors you have seen in BSim have rough analogs programmed into RoCK's firmware.

RoCK is based on Atmel's versatile and powerful AVR AT90S8535 microcontroller. Built into the 8535 are analog to digital converters (ADCs), a comparitor, and three timer/counters. The microcontroller also contains 8kB of FLASH program storage, 512B of SRAM, and 512B of software-programmable EEPROM. The chip uses the "Harvard" architecture with separate memory spaces for data and program storage. We programmed RoCK in the C language (plus a small amount assembly code) using the IICAVR cross compiler from ImageCraft.

---

[2]The robots that RoCK enables are in many ways reminiscent of the Braitenberg vehicles illustrated in Figure 1.10.

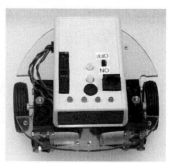

Figure 8.2

RoCK can be used to control a variety of mobility bases. On the left, RoCK's circuit board has been attached directly to a drive/steer base. The center photo shows the RoCK module attached to a differentially-steered tracked base. On the right, a home-brew chassis made from polycarbonate and motors scavenged from an old RC car forms RoCK's mobility base.

Burned into RoCK's FLASH-based firmware are 10 preprogrammed tasks plus two utility tasks. Users can program and store up to four additional tasks of their own. When a user presses the select button and twists the user potentiometer, RoCK can be commanded to perform any one of these 16 total tasks. RoCK's piezoelectric beeper can play user-programmed tunes of over 100 notes. Robot trajectories composed of more than 100 path steps can be stored and executed. RoCK has sensors to detect light, obstacles, collisions, and battery voltage. Connecting RoCK to a host computer enables the user to monitor the robot's sensors and to watch and set the robot's parameters. Despite its high level of functionality, RoCK remains an inexpensive device.

RoCK can be connected to a host computer for programming and monitoring, but the on-board user interface enables RoCK to run without being wired to a host computer. A power switch controls both the nine-volt logic supply incorporated into the RoCK module and the motor battery supply attached to the mobility base. A block diagram of RoCK's systems is shown in **Figure 8.3**.

The illumination pattern on RoCK's four LEDs indicates robot status. During task execution, LEDs generally display the state of the left and right obstacle detectors and the status of the colli-

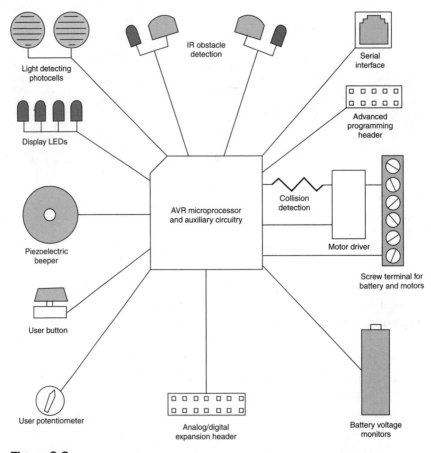

Figure 8.3

This functional diagram graphically depicts key elements of RoCK's hardware. RoCK's sensors include dual photocells, dual IR obstacle detectors, motor-stalled-based collision detection, and battery voltage monitors. The user interface consists of two inputs: the user button and user potentiometer, and two output devices: four status LEDs and a piezoelectric buzzer. RoCK can be connected to a host computer for purposes of monitoring and programming through a serial line and advanced programming header. RoCK can control two drive motors connected via the screw terminal.

sion detector. During task selection, the LED pattern indicates which of the built-in tasks the user has chosen.

The piezoelectric beeper can produce tones of arbitrary frequency, enabling RoCK to play built-in or user-programmed tunes.

Additionally, the state of certain sensors can be mapped into the frequency played by the beeper.

The user potentiometer enables the user to experiment with the parameters that control the robot's behaviors. For example, in a user task, the gain parameter in the control loop that enables the robot to follow a bright light can be mapped to the user potentiometer. The user can then adjust the robot's light-following response from sluggish to neurotic. The user button works in conjunction with the user potentiometer to select the built-in tasks the user wishes to run.

RoCK incorporates a dual-channel IR obstacle detection system. This system is composed of two series-connected emitters and two independently wired IR receivers. The receivers are sensitive to a 38-kHz modulation frequency. The emitter/detector pairs point diagonally outward from RoCK in such a way as to cover the area in front of the robot. Each receiver detects IR radiation reflected from nearby objects in the direction the detector is pointed, informing the robot of imminent collisions.

Dual diagonally mounted photocells measure the light level in front of RoCK. The difference between the light measured by the left and right photocells enables RoCK to home on a bright light source or speed to a dark corner.

As the robot operates, the logic and motor batteries discharge, decreasing the voltage of each battery. RoCK monitors the voltage of these two batteries, allowing the robot to take action if the voltage falls too low.

As we have seen in earlier chapters, an instrumented bumper provides an important means of collision detection. The mechanics of such a system are, however, rather complicated. RoCK avoids that complexity by using only a motor-stalled-based, one-bit collision detector. A calibration potentiometer is provided to adjust the trip point of this sensor to a value appropriate to the particular motors to which RoCK is attached.

Programming Specifications

A block diagram of RoCK's software architecture is presented in **Figure 8.4**. Using data provided by the task boxes on the left, the Task selector configures the behavior priorities used by the arbiter. The Task selector can also change the value of the various behavior parameters. The beeper is controlled in a way specified by the particular task. The User Tasks (the tasks programmed by the user) operate in a way identical to the other tasks, except that their specifications are stored in EEPROM, rather than flash memory. The scheduler runs one behavior after another and keeps the sensor values updated.

In developing RoCK, we first chose the set of tasks we wished the robot to perform and then created the primitive behaviors to support those tasks. (**Table 8.1** outlines RoCK's tasks.) In the code, we specify a particular task by choosing a list of behaviors, ordering their priorities, and choosing values for behavior-relat-

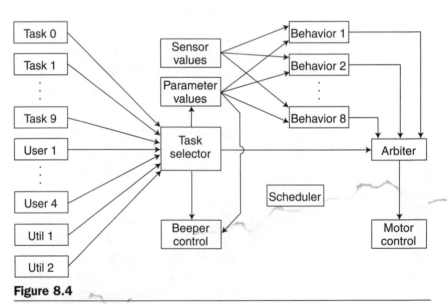

Figure 8.4

This simplified diagram shows the structure of RoCK's software. The diagram includes the familiar set of behaviors connected to an arbiter, but adds some functionality in the form of the Task selector. The Task selector makes it possible for the user to command the robot to run different tasks. The scheduler provides the "main loop" that in turn runs all the other pieces of software.

ed parameters. The reusability of these components enables us to store many tasks in the robot.

Table 8.1   These 16 named tasks can be selected via RoCK's on-board user interface. The left-most column indicates the number displayed in binary by RoCK's LEDs during task selection.

LED	Task	Description
0	Theremin	The Theremin task is the only task that does not make the robot move. This task uses the photocells and the beeper to simulate the Theremin, an early electronic musical instrument. The difference in the light level falling on the left and right photocells controls the frequency of sound emitted by the beeper.
1	Waltz	Waltz makes the robot move in a programmed pattern while a tune plays on the beeper. The user potentiometer controls robot speed. Like most tasks, Waltz specifies the Escape behavior as its highest priority behavior. This means that if the robot bumps into something while Waltzing, the robot will attempt to extricate itself before continuing with the dance.
2	Wimp	The Wimp task gives the robot a shy personality. The robot sits motionless until an object moves close enough to be sensed by the IR detectors. When this happens, the robot backs away.
3	Schizoid	Schizoid gives the robot a nervous personality. The robot cruises in a straight line, occasionally spinning or turning randomly. The interval between these events is controlled by a parameter. Increasing the parameter makes the robot seem calmer; decreasing the parameter makes the robot more frantic.
4	Pounce	The Pounce task has the robot sit in one spot until something comes near enough to trip the IR

continued on next page

Table 8.1 These 16 named tasks can be selected via RoCK's on-board user interface. The left-most column indicates the number displayed in binary by RoCK's LEDs during task selection. *(Continued)*

LED	Task	Description
4	Pounce	detectors. The robot then races full speed forward until it collides with the encroaching object.
5	Moth	The robot uses the difference in the left and right photocells to servo toward the brightest light. The robot will thus respond as if it were a moth homing on a flame.
6	Mouse	The Mouse task enables the robot to follow walls. Using its IR detectors, the robot cruises along a wall, going through doorways and turning away from inside corners.
7	Chicken	In this task, the robot uses its IR sensors to play chicken. The robot travels along at high speed, turning away just before colliding with objects it encounters.
8	Roach	When the room is dark, RoCK waits patiently. But when the light is switched on, RoCK races toward a dark spot. Safely concealed in the shadows, RoCK comes to a halt.
9	Remote	The Remote task allows direct control of the robot by the user when the robot is connected to the host computer. Remote is implemented using only the Joystick behavior. Thus the user is responsible for preventing collisions.
10	User 1	User-programmable task number 1
11	User 2	User-programmable task number 2
12	User 3	User-programmable task number 3
13	User 4	User-programmable task number 4
14	DiffSel	Helps the user correctly connect a differential drive base
15	SteerSel	Helps the user correctly connect a drive/steer base

RoCK's Behaviors

All twelve of RoCK's built-in tasks (numbers 0 through 9 plus 14 and 15, as listed in **Table 8.1**) and all four of RoCK's user-programmable tasks can be constructed from the set of eight primitive behaviors listed in **Table 8.2**.

Table 8.2   Each of RoCK's primitive behaviors can have a number of parameters that affect the behavior's output. Most parameters are unique to a particular behavior; the global speed parameter, however, is used by a number of behaviors.

Index	Behavior	Parameters
1	Dance	dance_tune_index, tempo, speed
2	IR_follow	speed, nine gain parameters (see text)
3	VL_follow	speed, nine gain parameters (see text)
4	Boston	time_between_events, event_duration,
5	Cruise	speed, turn_angle
6	Escape	backup_dist, spin_dist
7	Joystick	turn_angle, speed
8	Wire	

Dance

The Dance behavior causes the robot to move according to a stored program. A dance is a sequence of motion commands that specify a relative turn angle and the duration of motion at that angle. A motion command is stored in a single byte using the format: [tttfffff], where ttt is three bits of duration information and fffff is five bits of angle information. Dance has one parameter, dist_factor. For each motion step, Dance multiplies dist_factor by the duration bits to compute the length of time the robot holds the associate turn angle. One dance program is built into RoCK, but the user can program and store into EEPROM a second dance of over 100 steps.

IR_follow and VL_follow

The IR_follow and VL_follow behaviors are very similar in operation. The former computes robot motion using data from the left and right IR sensors; the latter uses information from the left and right photocells. These motion computations are made using a very general technique that provides an arbitrary linear mapping from a two-degree-of-freedom (DOF) input into a two DOF output. The technique also computes whether to request control of the robot. Both behaviors work in a manner fully analogous to the general linear transform we saw in the section, "Generalized Differential Response," in Chapter 5.

Boston

Periodically, the Boston behavior causes the robot to swerve randomly. Most of the time, Boston outputs no motion commands. Occasionally, however, Boston specifies an arc, rotation, or other motion. The parameter time_between_events specifies the number of seconds between such events. The event_duration parameter determines how long each motion event lasts. A random factor is included in the motion computation to make the behavior less regular.

Cruise

Cruise ignores all sensors and makes the robot move at a constant velocity. The behavior has one parameter pc_dir—the desired relative turn angle of the robot. Cruise is explained in more detail in subsection "Behavior Format" later in this chapter.

Escape

The Escape behavior attempts to extricate the robot after it has collided with an object. When the collision detector triggers, Escape responds by commanding the robot to back up and spin in place before releasing control. The duration of each of these steps is controlled by a parameter.

Joystick

Joystick allows the user directly to control the robot when the serial cable is connected. Robot speed and heading are obtained directly from the host computer.

Wire

The Wire behavior is a utility behavior included as a debugging tool for the user. Wire activates RoCK's LEDs and the mobility base's motors in a regular way that reveals to the user whether the mobility base is connected to RoCK in the correct way.

Beeper Control

Rather than build a second arbitration method for the beeper, we require that each task specify one of a small number of ways of controlling the beeper. The defined methods for controlling the beeper along with their descriptions are:

1. **Flash_tune**—The beeper plays a tune stored in flash memory.

2. **EE_tune**—The beeper plays a tune that the user composes and stores in EEPROM.

3. **Photocell difference**—The beeper behaves like a theremin—the beeper's output frequency is a function of the difference in light intensity sensed by the two photocells.

4. **Photocell sum**—Beeper frequency is a function of the total light intensity: the brighter the light, the higher the frequency.

5. **Bumper**—The beeper beeps in monotone when a collision occurs.

6. **IR_detect**—The beeper frequency depends on the IR detector: no sound for no detection, and different tones for detections by the left, the right, and both detectors.

7. **None**—The beeper is silent.

The Code

RoCK attempts to pack a great deal of functionality into a small code space. The robot has neither the room nor the need for an operating system. Given that we are flying solo, as it were, the first choice to make is the type of scheduler.

We think of all RoCK's software components (shown in **Figure 8.4**) as running in parallel. If this were actually the case, there would be no need for a scheduler. The scheduler creates the appearance of parallelism on a microprocessor whose operation is fundamentally serial.

RoCK implements a type of parallelism known as cooperative multitasking. Cooperative multitasking makes the scheduler's job especially easy. In this scheme, each parallel element (each behavior, the Arbiter, the utility functions that update sensor values, and so on) runs for a brief time, then returns control to the scheduler. The scheduler then calls the next element. Thus RoCK's scheduler consists of a list of subroutines that are called endlessly one after another.

The low overhead of a cooperative multitasking system makes it expeditious. On average, RoCK gives each behavior a chance to run over 500 times each second! One downside to cooperative multitasking is the discipline it requires of the programmer. Each element must be designed to compute for only a short time, then return control to the scheduler. A behavior that runs too long will freeze out all the other behaviors.

Scheduler

Here is RoCK's cooperative scheduler:

```
void main(void) {   // The scheduler is implemented by the main function
   // Decelerations
   extern unsigned char sensor[];
   extern unsigned char winner;        //Stores the ID the winning behavior
   extern unsigned char motor_bat_OK;
   // Initializations
   init_multi();
```

```
init_params_flash(); // Initialize default parameter values
init_serial();          // Connection with host
drive_config = EEPROMread(DRVSTR_CONFIG_ADDR); //Stored drive config
task_index = EEPROMread(TASK_INDEX_ADDR);
init_tasks(task_index);  // On startup pick the task to run
SEI();                      // Enable interrupts
// Tell the host where the sensor and parameter arrays are in RAM
//             EEPROM address        EEPROM data
EEPROMwrite(EE_SENSOR_ADDR_H, SENSOR_ADDR_H);
EEPROMwrite(EE_SENSOR_ADDR_L, SENSOR_ADDR_L);
EEPROMwrite(EE_PARAMS_ADDR_H, PARAMS_ADDR_H);
EEPROMwrite(EE_PARAMS_ADDR_L, PARAMS_ADDR_L);
wakeup();                 // Play wakeup tune and show behavior selection
//The Main Loop — Read sensors, run behaviors, and arbitrate
while(1) {
  acquire_sensors();              // Read all analog and digital sensors
  p_arr[px_frob].u = user_pot;    // Move data to selected parameter
  eewriter();        // Maybe host wants to write to EEPROM
  choose_task (); // Maybe user wants to pick a different task
  buzzer_beh(pb_select);  // The "behavior" that controls the buzzer
  // Give each behavior a chance to run:
  cruise();          // Move at a constant selected velocity
  joystick();        // Let the user control the robot
  IR_follow();       // IR_follow behavior
  vl_follow();       // Visible Light follow behavior
  escape();          // Escape from bump behavior
  boston();          // Randomly do something
  dance();           // Move to music
  wire();            // Maybe help user to connect the motors
  winner = arbitrate();   // Find highest priority beh that wants ctl
  // LED control
  if (choose_halt)       // If user is choosing a task
    leds(led_var);       //   Display the current choice
  else if (winner < WIRE_ID) {        // The normal situation
    led(0x01, (IR_detect & IR_LEFT )); // Left IR sees an object
    led(0x08, (IR_detect & IR_RIGHT)); // Right IR sees an object
    led(0x02, !motor_bat_OK);          // Glow if motor bat too low
    led(0x04, bump);                   // Declared bump
```

215

```
    }
    move_dual(winner);                // Dif drive or drive steer
}}
```

A lot is going on in this section of code, but the key item to note is the while-loop that contains calls to each of the primitive behaviors. Each iteration of this loop begins with a call to the subroutine that gathers sensor data. Subsequently, each behavior is called in turn and given a chance to run. The fact that the behaviors run in the order shown in the code has no bearing on behavior priority. After all the behaviors have been called, the arbiter runs and picks the winning behavior. The last statement in the loop is the call to move_dual. This routine sends the motion command computed by the winning behavior to the robot's motors. Nothing especially mysterious or magic is taking place.

Behavior Format

Next we'll take a look at the code that implements a couple of RoCK's behaviors. We begin with the famously simple Cruise behavior. Cruise is actually written using three macros, but to avoid confusion, let's look first at Cruise in its expanded form[3]:

```
void cruise(void) {
    unsigned char behavior_id = CRUISE_ID;
    _drive_angle(pc_dir, pg_speed, behavior_id);
    beh_ctl[behavior_id] = behavior_id; }
```

Cruise is composed of only four lines of code, but analyzing this little bit of software will take us through most of the complexity in RoCK's behavior-based implementation.

---

[3]In C, a macro allows us to substitute one string of text for another. When the code is compiled, the macro text is replaced (we say the macro is expanded) by the text in the macro definition. Macros are especially useful when a bit of code has to be repeated in many places with small variations at each repetition. Used in this way, macros simplify your code and reduce the likelihood of certain types of errors.

The first line creates a subroutine named cruise and declares that it has no arguments. This is expected because, as we saw in the scheduler, behaviors are implemented as subroutines. (We will designate as Cruise the behavior implemented by the cruise subroutine.)

The second line of the subroutine creates a local variable called behavior_id and assigns the value CRUISE_ID to that variable. Each behavior in RoCK must have its own unique identifier. In one of the files that compose the RoCK system, numbers are attached to behavior identifiers in this way:

```
// Behavior IDs
#define STOP_ID        0    // ID of default stop behavior
#define DANCE_ID       1    // Dance a canned dance
#define IR_FOLLOW_ID   2    // Use IRs to follow/avoid objects
#define VL_FOLLOW_ID   3    // Follow/avoid visible light
#define BOSTON_ID      4    // Do random motion at random time
#define CRUISE_ID      5    // Move at a constant speed/rotation
#define ESCAPE_ID      6    // Respond to collisions
#define JOYSTICK_ID    7    // Direct control by user
#define WIRE_ID        8    // Help the user connect motor wires
#define MAX_BEH        9    // One more than the highest behavior ID
```

Each #define statement introduces a new macro. The macro tells the compiler to replace the first text string after #define with the second string of text wherever the first string occurs in the user's program. The replacements occur before the code is processed further. Thus in the Cruise behavior, CRUISE_ID is replaced by 5 so that the local variable behavior_id is assigned the value 5. (Text following the "//" is marked as a comment and ignored by the compiler.)

So far so good, but why does each behavior need a local variable called behavior_id and why does each instantiation of that variable need a unique value? This is done because RoCK depends on these features to enable arbitration.

Every behavior computes motor control commands for the robot. In RoCK, motor control commands are stored in the form of a left motor velocity and a right motor velocity. But clearly, behaviors must not write the value of the left and right motor velocity directly to the motor controller, because this would bypass arbitration and lead to contention between behaviors. Instead, each behavior that wants to control the motors writes its command to a particular slot in a pair of special arrays. The arrays are defined in this way:

```
int right_vel[MAX_BEH];        // Velocity array for right drive motor

int left_vel[MAX_BEH];         // Velocity array for left drive motor
```

And which slot of the right and left velocity arrays does a particular behavior write to? It writes to the slot specified by behavior_id. Each behavior has a unique local value for this variable, so that each behavior writes to a unique slot of the velocity arrays. Thus no behavior overwrites the commands of any other behavior.

The next line, line 3, of the Cruise behavior is:

```
_drive_angle(pc_dir, pg_speed, behavior_id);
```

This statement (ultimately) causes the left and right motor velocity commands to be written to the velocity arrays using behavior_id as the index. But where in this statement are the left and right velocities? There are many equivalent ways to specify the motion of a robot (see Appendix A). In this case, _drive_angle gives us a way to specify a relative heading and a velocity. The function interprets its first argument as an angle and the second as a velocity—both are parameters. The angle and velocity arguments are converted by the function into left and right wheel velocities and written into the corresponding velocity array. The angle argument, pc_dir, tells the function how tightly to turn the robot. When this argument is 0, the robot moves straight ahead— the velocities of the left and right wheel motors are both equal to pg_speed. When the angle argument is 180 degrees, the robot moves backward. When the argument is 90 degrees, the robot spins in place to the left; –90 makes the robot spin to the right, and so on.

The final statement in the Cruise behavior is:

```
beh_ctl[behavior_id] = behavior_id;
```

By this point, the behavior has computed motor commands for the robot. But is Cruise triggered or untriggered? Should the arbiter pass to the motor controller the velocity commands that Cruise has computed or not? By writing its own (non-zero) identification number to the beh_ctl array, Cruise informs the arbiter that the motor commands should be delivered. Had Cruise not wanted control, it would have executed the statement:

```
beh_ctl[behavior_id] = 0;
```

Were you to look at the file that instantiates RoCK's Cruise behavior, you would find it written not as shown above, but rather in this way:

```
void cruise(void) {
  behavior(CRUISE_ID);               // Declare behavior_id
  drive_angle(pc_dir,pg_speed);
  CONTROL; }
```

This version operates exactly as described above, but here we have used three macros to hide the details of the implementation. The second line automatically creates local variable behavior_id and assigns to it the value CRUISE_ID. The third line is a macro that expands to the _drive_angle function. The final line becomes beh_ctl[behavior_id] = behavior_id;

We'll take a quick look at one more behavior, the Boston behavior. This behavior, implemented as an FSM, causes the robot to make a random motion at a random time.

```
void boston(void) {
  behavior(BOSTON_ID);
  static int angle;
```

```
static unsigned char state = 0;       //The state of the FSM
extern unsigned int  clock;
static unsigned int  del_time;
static unsigned int  b_time;
switch(state) {
  // State 0 - Compute time until random event
  case 0:
    b_time = clock;                       // Starting time
    del_time = ((rand()>>10) * (unsigned int) pr_time);
    PUNT;                                 // Don't try to control robot
    state++;                              // Go to next state
    break;
  // State 1 - Wait for event to start
  case 1:
    if (time_out(b_time, del_time)) {     //If done act, else remain
      b_time = clock;                     // New timeout time
      del_time = ((rand()>>7) * (unsigned int) pr_dur); //
      angle = rand()>>7;                  // Pick a random angle
      drive_angle(angle,pg_speed);        //Macro hides array details
      CONTROL;                            // Try to control robot
      state++;                            // Go to next state
    }
    break;
  // State 2 - Random heading until timeout
  default:
    if (time_out(b_time, del_time))
      state = 0;
    break;
  }
}
```

Boston has the same basic format as Cruise. The statement, behavior(BOSTON_ID), appears early in the subroutine to give Boston its unique identifier and declare the behavior_id local variable. Second, the code calls drive_angle in various places when it wants to send velocity commands to the motors. And third, PUNT and CONTROL are called to inform the arbiter of Boston's desires. (PUNT writes a zero into Boston's slot in the beh_ctl array.) Boston uses C's switch function to implement an FSM. In state zero, Boston figures out how long to wait until the next random motion, then advances to state one. Boston waits in state one until it is time for the event to occur. At the proper moment, the random motion is commanded, the time when the motion should stop is computed, and control passes to the next state. Boston remains in the final state, state two, until the event finishes; then control goes back to state zero.

Here is one final example of a RoCK behavior—RoCK's version of Escape:

```
// Escape – A ballistic behavior executed when a collision
// is detected. The robot will backup and turn before releasing
// control. Parameters in units of ~mS *6 => max time = 256 *0.016
// => 4.096 seconds. That is, each unit of pe_x is 16 mS.
// Use motor current to determine a bump. Backup and spin if detected.
void escape(void) {
  behavior(ESCAPE_ID);                    // Declare behavior_id
  extern unsigned char sensor[];
  static unsigned char state;
  extern unsigned int  clock;
  static unsigned int  e_time;
  static unsigned int  duration;
  switch(state) {
    case 0: if (bump) {                   // There is now a collision
              e_time = clock;
              duration = ((unsigned int) pe_backup) << 4;
              drive(-pg_speed,-pg_speed);// Backup fast
              state++; }                  // Advance to next state[4]
            else
              PUNT;                        // No bump => release
            break;
    case 1: if (time_out(e_time, duration)) {
              e_time = clock;
```

---

[4]Why is there no CONTROL statement in this section of code? None is needed here because on the next iteration, the behavior will be in the next state (case 2) and this state does contain a CONTROL statement.

```
            duration = ((unsigned int) pe_spin) << 4;
            drive(-pg_speed,pg_speed);  // Spin in place
            state++; }
      CONTROL;
      break;
   default: if (time_out(e_time, duration))
            state = 0;
      CONTROL;
      break;
   }
}
```

Most structural aspects of this behavior should now be familiar. The RoCK robot does not have a differential bumper; thus RoCK can detect only that a collision has occurred, it does not know on which side of the robot the collision happened. Escape determines that a bump has occurred by examining the global variable bump (whose value is either zero or one). Also, RoCK lacks shaft encoders, meaning that it cannot determine how far it has moved. All that RoCK can know is that it commanded a certain velocity and that enough time has passed to allow the robot to move the desired distance. Escape therefore decides that it has backed up and spun far enough based on time parameters pe_backup and pe_spin.

Arbiter

RoCK refers to arbiter by the name arbitrate. There are two features peculiar to RoCK that make arbitrate a little more complex than it might otherwise be. RoCK is not programmed to do just do one task; it can perform any of 16 different tasks. A task may use some, but not other behaviors. Each task can prioritize the primitive behaviors that it does use in any way. A task can also set global parameters and parameters associated with the primitive behaviors in an arbitrary way. A second complication of RoCK, compared to robots that do just one task, is that the user's selection of a new task is treated outside the arbitration system.

RoCK's arbiter is defined in this way:

```
int arbitrate(void) {
  unsigned char pri_index;      // Check each index
  unsigned char beh_index;      // Get the behavior index stored there
  // Step through the behaviors in the beh_pri array
  if (choose_halt)              //Don't arbitrate if user is choosing a new task
    return 0;                   // This forces the STOP "behavior" to win
  else {
    for(pri_index = 0; pri_index < MAX_BEH; pri_index++) {
      beh_index = beh_pri[pri_index];      // Get slot # of next hst beh
      if (beh_ctl[beh_index])              // If this behavior wants control
        return beh_index;                  // ...return it as winner
    }
    return 0;              // No behavior wants control, so STOP
  }
}
```

Unless the user is in the process of selecting a new task, the "for" loop runs. This loop steps through the possible priorities. If there are *n* primitive behaviors, then there must necessarily be *n* different priorities. Were there fewer priorities than behaviors, then two or more behaviors could have the same priority—this would lead to an ambiguity.

Arbitrate uses two levels of indirection,[5] as shown in **Figure 8.5**. The beh_pri array contains the identification numbers of the primitive behaviors that implement a particular task; these numbers are ordered according to index number. That is, the identifier of the highest priority behavior is stored in beh_pri[0], the identifier of the second highest priority behavior is stored in beh_pri[1], and so on.

Arbitrate steps through the elements of beh_pri one at a time, assigning the identifier to the variable beh_index. Arbitrate then uses beh_index as an index for examining the beh_ctl array. If the value of beh_ctl[beh_index] is zero, then the behavior correspond-

---

[5]An indirect reference means that the value we are interested in is not stored in the referenced location. Rather, stored at that location is a pointer to another location, where the value (or in RoCK's case another pointer) is stored. Indirection can make programming structures more versatile.

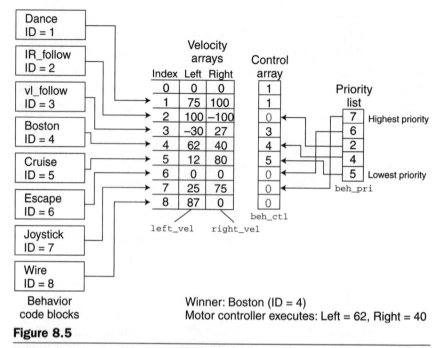

Figure 8.5

To instantiate a task, the identification numbers of the behaviors that compose the task are written into the beh_pri array. The behavior whose identification number is stored in index zero has the highest priority in this task; the behavior whose identification number is stored in index one of the beh_pri array has the second highest priority, and so on. The arbiter steps through the elements of the priority list using the numbers stored there to index into the control array, beh_ctl. Any behavior wanting control (any triggered behavior) writes its ID number into its slot of the control array. In the example, the Boston behavior has ID=4. Boston, wanting control, writes 4 into slot 4. No higher-priority behavior (Joystick, Escape, or IR_follow) is requesting control, so the arbiter ultimately checks the priority list slot containing the fourth highest priority. Finding a 4 stored in this slot, the arbiter notes that slot 4 of the control array is non-zero and thus declares Boston the arbitration winner.

ing to the identifier stored in beh_index does not want control and the "for" loop continues iterating. If the value of beh_ctl[beh_index] is non-zero, then the corresponding behavior does want control and the loop is exited with the index of the highest-priority behavior that does want control (the winning behavior) stored in beh_index. Arbitrate returns this value.

In RoCK, a small set of data represents a task. The set contains information for ordering the priorities of the primitive behaviors, values for relevant behavior parameters, and a mapping from the user potentiometer to one of the parameters (this gives the user real-time control over one parameter). The code that specifies RoCK's built-in behaviors is contained within an initialization routine; data for instantiating user-programmed behaviors is burned into EEPROM. The following block of code from RoCK's startup routine illustrates how built-in tasks are instantiated:

```
// Schizoid
    case 3: beh_pri[0] = ESCAPE_ID;    // Highest priority beh
        beh_pri[1] = BOSTON_ID;
        beh_pri[2] = IR_FOLLOW_ID;
        beh_pri[3] = CRUISE_ID;        // Lowest priority beh
        pi_a = -64; pi_b = -64; pi_c = 64;
        pi_d = 0;   pi_e = 0;   pi_f = 64;
        pi_g = 1;   pi_h = 1;   pi_i = 0;
        px_frob = ig_speed;            // User pot controls robot speed
        pb_select = BUZZ_PHO_DIF;  // Theremin makes it sound weird
        pr_time = 32;                  // Parameters
        pr_dur = 2;
    break;
```

Seventeen assignment statements instantiate the Schizoid task. The highest priority primitive behavior in Schizoid is Escape. Assigning Escape_ID to the zeroth slot of the priority array beh_pri[0] gives this behavior the highest priority. The other behaviors, in order of priority, are Boston, IR_follow, and Cruise. Variables that begin with a 'p' and contain an underscore are behavior parameters. Schizoid uses the general linear transform method to control the robot's response to the IR sensors. The nine variables pi_a through pi_i are the matrix elements. The other parameter settings control other aspects of the task.

The advantage of building RoCK's system as outlined here is that additions and deletions are very simple—just storing some behavior identifiers into an array and writing some values into some variables creates an all-new task. Adding tasks, deleting tasks, and modifying tasks are all very simple—just as with BSim.

Summary

This chapter has demonstrated one method for implementing a behavior-based software system on an inexpensive, low-powered microcontroller. Is the system presented an optimal example? Indeed it is not. It is just one way that such a system might be implemented and far from the best way. There are, no doubt, simpler systems that accomplish the same things. It would require but little programming prowess to best the author in this matter and design a superior system of your own.

The only virtues of the example presented are that it is true to the principles of the behavior-based programming paradigm and it demonstrates elements you can expect to find in any behavior-based programming system. In particular, when you design such a system, you will need to implement:

- A scheduler

- A behavior format

- A way to specify priority

- A connection method

- An arbiter

Cooperative multitasking makes the scheduler trivial—it's just a loop of subroutine calls. Behaviors can be subroutines—they must have the ability to maintain state from call to call. You must develop a way to make connectable inputs and outputs explicit and you must resist the temptation to abuse the connections. Assigning a unique number to each behavior can specify priority. Finally, you must build an arbiter that can examine the priorities of each behavior and pick a winner.

Exercises

Exercise 8.1 The behaviors in RoCK are called in a particular order by the scheduler—Cruise is first, Joystick second, and so on. Does the order have any importance? What would happen if the order of calling behaviors changed from iteration to iteration?

Exercise 8.2 Suppose you want to add one more behavior to the RoCK system. List the steps that you must take to accomplish this.

Exercise 8.3 List the steps that would be required to add a new task.

Exercise 8.4 How many elements must RoCK's beh_ctl array have?

Exercise 8.5 What determines the number of elements the array beh_pri must have?

Exercise 8.6 Must behaviors in RoCK explicitly give up control or do they lose control automatically after a certain time?

Exercise 8.7 What will be the effect on the system if a particular RoCK behavior computes for, say, one second before returning control to the scheduler?

Exercise 8.8 If, during a particular iteration, no behavior requests control, what will be the contents of the beh_ctl array?

Exercise 8.9 Given the example shown in **Figure 8.5**, what does the robot do if none of the behaviors requests control?

Exercise 8.10 Is there any significance to the order in which the behavior code blocks are listed in **Figure 8.5**? Is there any significance to the ID numbers each behavior is assigned?

Exercise 8.11 Describe the task implemented by the example in **Figure 8.5**.

Exercise 8.12 What happens if the user tries to drive the robot into a wall, given the task configuration of **Figure 8.5**? How would you change the task specification to enable the robot to protect itself?

Exercise 8.13 Suppose that RoCK were designed to perform only one task that could not be changed. In what ways might the system be simplified? In particular, would there be a need for the beh_pri array?

Exercise 8.14 Write a new behavior, Bright_freeze, that makes the robot halt any time the total light measured by the photocells exceeds the value L_0. The behavior ID for Bright_freeze is BF_ID. Assume that global variables left_photo and right_photo are provided and constantly updated by the system. If you are familiar with the language, write the behavior in C, otherwise use pseudocode.

Exercise 8.15   Write a new behavior, Find_light, that triggers when the total light level falls below L_0. When this happens, the robot should spiral outward until the light level rises above L_0. As a second step, add hysteresis to prevent the robot from entering and leaving this behavior too rapidly when the light level is close to L_0.

9

Future Robots

Fortune smiled on me in 1982. My new job at the MIT AI Lab and my enduring fascination with robots both began that year. In 1982, working with robots in any capacity demanded good fortune, because only a scant handful of institutions were then engaged in robotics research. High cost and the need for advanced technical expertise put robotics out of reach for most universities, not to mention most individuals. Starting a robotic project in those days meant spending thousands of dollars for computers and sensors and devoting perhaps hundreds of hours to machining parts and building equipment. Robotics was an enterprise reserved for the well-funded elite—high school students and college undergraduates need not apply.

Today life has changed. Driven by the same economic and technological forces that make computers and consumer electronics ever more affordable, hobby and educational robots have become widely accessible. As a result, at every level, interest and participation in robotics is burgeoning. Courses, contests, and companies have come into being to teach, promote, and exploit robots. You no longer need a generous government grant and a PhD in electrical engineering, computer science, or mechanical engineering to build a robot. But building and programming your own robot might just induce you to pursue a PhD in one of those fields—and apply for a government grant!

At the current state of the art, robots offer tremendous opportunities for individual learning. Designing, constructing, and programming robots are challenges both engaging and enlightening. Each year more and more educators and students accept these challenges and reap rich intellectual rewards from their robotic explorations.

Reply Hazy—Ask Again Later

But where are robots heading? Beyond facilitating the education of interested individuals, beyond relieving the tedium of a couple of household chores, what can we reasonably expect from future robots?

For years, breathless reports of amazing, just-around-the-corner robots have tantalized the public. But in nearly every case it seems, these robots have accomplished their wonders only by virtue of the umbilical tethering them to the laboratory of their genesis. How can we sever this fetter? How do we transform robots into practical, useful devices seen not just in news stories but on store shelves? What robotic talents and skills will enable robots to make the leap?

Robots, it would seem, still have a long way to go. The most capable autonomous mobile robots commercially available today—robots that clean floors or mow grass—pale in comparison to the appealing but insubstantial robots of popular imagination. In compelling works of fiction, Isaac Asimov, George Lucas, and many others promised robots that could converse with us, bear our burdens, and protect us. Contrast these dreamed-of competencies with the capabilities autonomous mobile robots have demonstrated to date. Existing robots can avoid traps and tumbles, they can keep their batteries charged, and they can perform a couple of useful but rather simple tasks.

If the initial goal of behavior-based robotics had been to grant mobile robots insect-level intelligence, then mowing the grass and cleaning the floor is about the level of ability you might expect from a programmable insect. But this is no place to stop. Surely with a bit more insight, creativity, and hard work we can continue the process and devise yet more clever robots, ones able to rival the intellectual prowess of a lizard, for example.

With only the earliest of robotic capabilities now in place, much work remains undone and many puzzles linger unsolved. But the potential benefits of continued research and development are compelling. More than a few robotics practitioners foresee the day when robots will surpass the most hopeful visions of science fiction writers.

Beyond taking over undesirable jobs, robots will enable us to accomplish tasks that were never before possible. Humans come in only a limited range of sizes, possess a restricted set of low-force actuators, and have an incomplete sensor suite. Not so with robots. We can specialize a robot to the task we wish to perform. Perhaps we need a robot less than five inches tall that is able to lift 500 pounds, and sense ionizing radiation. We can potentially design a robot with these or any other beneficial combination of characteristics.

Maybe it's not too much to wish that a lizard-smart robot might acquire capabilities that push the envelope of what is now possible. Based on what we are able to do today, here are a few example abilities that seem not too far-fetched. An affordable, practical robot should:

- Reliably recognize places that it has visited before.

- After an excursion, reliably navigate back to its starting point using only on-board sensors (that is, independent of external aids like GPS).

- Cover an area systematically, rather than randomly.

- Differentiate between humans and objects or between humans and other animals.

- Dependably identify certain classes of objects (weeds, for example) in natural environments.

- Grasp and manipulate certain classes of natural objects (fruit, berries, or rocks, for example).

- Navigate autonomously in challenging outdoor terrain or rubble.

It is not sufficient for one-of-a-kind laboratory-based robots to exhibit capabilities such as these. The aim of the practical roboticist must be to instill such abilities into economically viable robotic products.

With these and a few other capabilities, we might call on robots to do such things as neutralize buried land mines, reduce or eliminate the need for pesticides and herbicides by having robots eradicate insect pests and weeds via mechanical means, search for lost hikers and missing children in difficult environments, locate victims buried by a collapsed building or an avalanche, or mine underground or undersea resources.

What hurdles obstruct the path from here to there? What capabilities are most in need of further refinement or invention?

Which Path?

An earnest and ongoing debate within the Artificial Intelligence community seeks to answer this question: What route should we follow toward creating more intelligent machines? One camp promotes a traditional approach that harks back to the founding of Artificial Intelligence as an independent field of study. The traditional method, sometimes called Good Old Fashioned Artificial Intelligence (GOFAI), can be seen as a top-down approach. Proponents argue that pure thought implemented through symbol manipulation is both necessary and sufficient to enable machine intelligence. Give a computer plenty of information, give it rules for processing that information, and at some point the computer will be able to think as we do—drawing conclusions based on what it knows and inventing reasonable new ideas. Intelligence, in this view, amounts to the clever manipulation of a large database of facts—no direct sensory or motor connection to the outside world is needed.

Another camp regards the connection of mind to sensing and actuation as absolutely indispensable to the development of intelligent machines. The goal is not so much to make machines think intelligently as to make them act in an intelligent way. This group takes a bottom-up approach, attempting to follow the pattern of evolution—start with simple robots of modest abilities, but insist that they inhabit the real world. Behavior-based robotics is congruent with this path.

Both methodologies have scored victories. And although my own bias is toward the bottom-up approach, clearly both points of view offer key insights. Perhaps in the case of robots, the ultimate venue for this contest of ideas will be the marketplace. Rather than survival of the fittest in the natural environment, the best procedure for building a robot may ultimately be resolved in favor of the method (or blend of methods) that can produce the most economically viable machines—survival of the thriftiest.

Where might an aspiring roboticist apply his or her talents to make the greatest difference in the years ahead? There are plenty of options; the next sections suggest some possibilities.

Future Actuation

Although the focus of this text is on programming, we have seen that robotic systems cannot yet be neatly separated into isolated modules. The sorts of programs that you can write are limited by the realities of actuation and sensing—to build effective robots, these boundaries too must be pushed back. For present purposes, we will let actuation stand in for all things mechanical and power related.

Power

Robotic actuators need power, but as suppliers of power batteries, they are abysmal. A key measure of battery functionality is energy density, the maximum number of watt-hours or joules that can be packed into a given weight or volume of battery. In the time it has taken computer memory chips to improve their storage density by a thousand-fold, the energy density of the best commonly available batteries has increased by only a few percent. A battery chemistry invented over 100 years ago (the LeClanche cell) is still available on store shelves today and still competitive with more recently developed chemistries.

For robots, the sorry state of energy storage technology makes everything harder. Small energy capacity means that robots have limited time to finish their tasks before power runs out. And that fact means that they must be efficient. But high efficiency is a lot to demand from a new and not-yet-mature technology. Alternatively, if we give a robot a battery large enough to accomplish a task of reasonable size, the robot becomes heavy and cumbersome. This makes the robot much more difficult to control and makes the consequences of mistakes (smashing into a person, for example) more severe.

Better power sources, be they advanced batteries, fuel cells, or micro-turbine generators, would make it far easier to design useful robots.

Motors

When a roboticist sits down to design the actuators of his or her robot, there aren't many choices. The most practical actuation source available is the permanent magnet direct current (PMDC) motor. This old workhorse has been around for a great many years and is perfectly adequate for a multitude of applications. But the PMDC motor leaves much to be desired from a robotics point of view. First, such motors spin too fast and deliver too little torque. To accommodate this deficiency, the robot builder invariably has to attach a gearbox to the motor to slow it down and "torque it up." This makes the total actuator bigger, heavier, and less reliable than it otherwise might be. Second, on a power-per-force or power-per-torque basis, an electric motor is unremarkable compared to muscle the material animal agents have to work with. Finally, small, inexpensive PMDC motors are also fairly inefficient. They can easily waste half or more of the power supplied to them. Such motors squander much of the limited energy available from the batteries as heat, rather than converting it into useful mechanical work.

A greater variety of motor types would make for better robots. Slow-turning, high-torque motors, high-efficiency linear motors, as well as many other sorts of motors would find ready application in robots.

Manipulation

To date, the most successful mobile robots are those that can accomplish their function primarily by rolling or otherwise moving around. The floor-cleaning robots move about, fully covering an area without getting stuck. And while they are doing this, such robots operate a cleaning mechanism that cleans the floor. Lawn-mowing robots move about the yard without getting stuck. And while they do so, they operate a cutting mechanism that trims the grass. Robotic spacecraft, robotic aircraft, and robotic underwater vehicles accomplish tasks that consist primarily of carrying a sensor package to some particular place or along a trajectory without crashing into anything.

But how many useful applications can be accomplished by nothing more than simply moving around? Often your first step in accomplishing a task is to move to some spot. Upon arriving at that spot, you pick something up, put something down, or otherwise alter objects or the local environment. Creatures across the animal spectrum do the same thing—first navigate, then manipulate.

Clearly, manipulation is a key ability for robots. Given the fact that assembly robots currently earn their keep in factories by positioning and assembling parts, it may seem that manipulation is a solved problem. It is not. Industrial assembly robots depend on being told in advance virtually their every move—operation in unknown and unstructured environments is problematic. Existing industrial manipulator robots are precise but unforgiving; they also tend to be heavy and power hungry. All of these features work to make commercial manipulators problematic for use on board small battery-powered mobile robots.

New thinking seems necessary. Manipulators that are small, lightweight, powerful, and energy efficient would fill the bill nicely. They should also be rugged, resistant to dirt, and, if they are to be used outdoors, waterproof and insensitive to temperature variations. One route to building manipulator systems for small mobile robots is to make the manipulator application-specific. That was the trick we used for our SodaBot in Chapter 6. This strategy also has evolutionary precedence—adaptation and specialization is the rule among living organisms. Build a mechanism to do one job well, rather than many jobs suboptimally.

In the animal kingdom, sensors and actuators are employed with abandon. Take a look at a blown-up photo of a spider in **Figure 9.1**

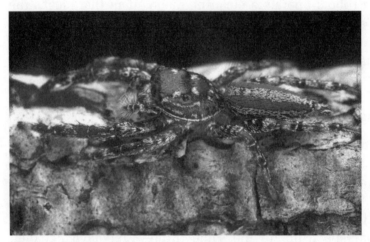

Figure 9.1

This jumping spider from Peru, like all spiders, is a marvel of sensing, actuation, and intelligent control. Spiders have eight legs with six joints per leg for a total of 48 degrees of freedom! Spiders have six or eight eyes; the retina of each eye may contain 800 individual sensory receptors. Thousands of tiny sensory hairs cover the surface of the spider's legs and body. Thus spiders have an order of magnitude, more actuators, and two or three orders of magnitude more sensors than any of the robots we have considered. According to some estimates, the processing and storage power of personal computers surpassed that of the spider brain in 1996.[1] As yet, however, no robot, regardless of the prowess of its processor, comes close to matching the abilities of this "simple" animal. *(Photograph courtesy of Chip Clark, Museum of Natural History, Smithsonian Institution.)*

---

[1]See *Robot: Mere Machine to Transcendent Mind* by Hans Moravec, Oxford University Press, 1999.

for example, and you will see many legs with many joints—far more than you will find on the most dexterous robot. A mobile robot designed to duplicate the mobility of the jumping spider would need 48 motors—likely making the robot too heavy to even shuffle along, let alone be able to leap to catch prey. Because of the unfavorable weight, torque, and power properties of PMDC motors, roboticists seek to minimize the use of motors in a design. So one answer to the challenge of building a better robot manipulator might "simply" be to design a low-cost motor more suited to the needs of robots. Another approach might be to find a way to achieve the functionality provided by many degrees of freedom while using only a few motors—on an application-specific basis.

Utilizing even more cleverness, one might design a manipulator with only limited degrees of freedom, but exactly the right degrees of freedom to accomplish some specific task.

Locomotion

Animals slither, crawl, walk, run, gallop, leap, climb, fly, glide, hover, paddle, swim, and dive—to name just a few locomotion methods. Commercial autonomous robots roll about on wheels and tracks and a few totter on mechanical legs. In order for robots to move out of the laboratory and other benign environments, robots will need to have better ways of getting around.

Consider such well-represented terrain types as a grassy plain, a leaf-littered forest, a thicket, a swamp, an ice floe, or a recently plowed farmer's field. There are many tasks we might like to have robots accomplish in such places, but how many existing robots could navigate autonomously in any of these terrain types? Before robots can be used in such places, many real-world mobility problems must be solved.

Future Intelligence

We've considered one basic methodology and many particular tricks for building the intelligence system of autonomous robots.

But what's missing? What other aspects of robot control are yet to be invented?

Robot Control

Behavior-based robotics is a robot-control paradigm that has shown great power and adaptability. It has been an enabling factor in robots that explore distant worlds and in robots that clean floors. But this method has had its most profound successes directing simple robots, robots that aspire to an intelligence level of, say, a spider. There are those who argue that the advantages of behavior-based programming will plateau at some point—the plateauing of exciting new methods has proved a near-universal occurrence in Artificial Intelligence.

We have lavished attention on each sensor and each actuator of the robots considered here—and done so to good effect. But we can be generous with our attention only because we have had so few sensors and actuators to consider. What happens when we build a robot that requires as many individual sensory receptors as a living creature—when the number of sensors climbs into the ones or tens of millions? Might new organizing principles be necessary?

At least part of the time, higher animals do seem to formulate and execute plans; we know that people do so. But planning fits somewhat uneasily into behavior-control programming. Will some mixture of GOFAI and behavior-based techniques provide the answer?

Learning

People learn throughout their lifetimes. We acquire new skills, new abilities, and new understanding. Sometimes learning is deliberate, but in many cases, it "just happens." With no conscious effort on the part of the learner, through repetition, our ability to accomplish some task improves. Thus learning seems an important component of all activity, from the highest level—studying physics, for example, to the lowest—the adaptations our muscle control systems make that enable better performance at, say, the trampoline. Robots, by comparison, are dunderheads.

Existing commercial mobile robots may respond in seemingly clever ways to hazards and opportunities in their environments, but they do not learn.

One subject area that especially interested me when I began working at the MIT Artificial Intelligence Laboratory was machine learning. I quizzed my boss on the subject, but he advised me not to bother trying to build learning mechanisms into a robot. He reasoned that I, the programmer, was much smarter than the robot. Given the state of the art at the time, I would always find it more effective to build a given ability into the robot directly, rather than have the robot acquire that ability on its own through learning. So far, no robot that I know of has proved the good professor wrong.

But this cannot be the last word. Ultimately, we need to have our robots learn from their experiences and indeed, research into machine learning has gone on for many years. Many specialized types of learning have been explored, including: adaptive control, reinforcement learning, genetic algorithms, and model-based learning. But there remains no established method that enables the general sort of learning of which people are capable. It is often said that you can learn only at the fringes of what you know already. Perhaps one reason that learning has proved so difficult to engineer in robots is because robots know almost nothing. As robots become more capable, devising ways for them to learn may become easier.

How do we add learning to the robot's repertoire of talents? Within the behavior-based systems we have studied, the most straightforward way for the robot to learn would be for it to write new behaviors of its own. Is there some mechanism that might make this possible?

Future Sensing

The greatest impediment to rapid robotic development, in my experience, is sensing. As stated earlier, if a robot can obtain unambiguous answers to the questions it must ask in the performance of some task, then it can successfully accomplish that task. The problem is that we can think of a great many tasks we would like robots to perform—tasks requiring answers that no sensor can currently give.

Vision

Throughout the animal kingdom, vision is present in myriad forms. Some organisms have an eyespot that may do nothing more than distinguish between light and dark. At the opposite extreme are birds of prey whose visual acuity exceeds our own. In between is a bizarre range of variations. There is, for example, the octopus. The octopus eye can tell the difference between horizontal and vertical lines, but cannot distinguish between a diagonal line that leans to the left and one that leans to the right! Spiders have different eyes for different purposes. Often several smaller eyes are used only for detecting motion; having detected motion, the spider orients to point its larger, imaging eyes at the moving object. Some spiders have eyes that they seem to use for measuring sky polarization; this can aid in navigation.

There was a time when vision researchers believed that the goal of their discipline was to find a way to transform the two-dimensional image from a video camera into a three-dimensional representation of the solid world. No more. There are a great many ways in which sensing the photons that happen to strike a creature's eye can be useful—finding a 3D representation of the world is only one of them.

In contrast to living creatures, nearly all mobile robots are effectively blind. But if the more-general vision problems are too difficult, perhaps, like other living creatures, robots can use vision in more application-specific ways. Finding robotic applications where some form of vision can be used in a specialized way seems a broad and fruitful area for exploration.

Acoustic Sensing

A flying bat can tell the precise direction, the size and speed of its prey, and maybe even what sort of insect it is about to encounter—all from a few sonar chirps. A robot can barely find an open, three-foot-wide door with its sonar sensors.

Try the experiment of walking around your home or office with your eyes closed while listening intently. (Be careful!) It is fairly easy to judge which room you are in just from the incidental sounds: the hum of the refrigerator, the muffling effects of a carpeted floor. By contrast, robots are deaf.

Clearly, there is a great deal more information in sound waves than robots currently extract. Developing capabilities based on passive acoustics or using active acoustics in some way other than conventional sonar appears to be another interesting area for further research.

Other Sensors

A bloodhound can find and follow the trail left by a person long after that person has passed by. The dog does this by detecting almost infinitesimal concentrations of odors characteristic of the particular person of interest—even when similar, competing scents are also present. The olfactory apparatus that makes this extraordinary feat possible fits in a tiny volume and consumes minuscule amounts of power. What might robots do if we could provide them with sensors that work as well? If such sensors were available, robots might help locate lost children, detect buried or hidden explosives, and perform other urgent tasks.

Fish possess a lateral line, a type of sensor that has no direct parallel in human experience. The lateral line, among other things, alerts the fish to the presence of other fish via minuscule disturbances in the water. It is easy to imagine providing robots with sensors analogous to human sensors. Conceiving of sensors that measure simple physical quantities like, say, the level of radiation or the presence of methane is also within our cognitive grasp. But might there be other odd but useful sensor systems like the fish's lateral line? Perhaps development of such sensors goes wanting not because they are difficult or expensive to build, but because nothing in our experience leads us to consider building such devices.

Robots could ask many more interesting questions—and thus perform many more constructive tasks—if only we had sensors able to answer those questions. Robots cannot find buried avalanche victims, or pull up weeds in a field, or locate deadly land mines without small, portable, affordable sensors able to detect such items. Creating new sensors with interesting properties is a very fertile field of endeavor that can be of tremendous benefit to robotics.

Exercise

Although we've nearly reached this book's final page, the story of robotics is far from complete. Writing the next chapter of that story is left as an exercise for the reader. Good luck.

A
Mathematics of Differential Drive

The programs that control autonomous mobile robots deal largely with the motion of the robot. Thus, a clear mathematical description of that motion is often essential for effective control. This and the problems inherent in a mathematical treatment of robot motion are discussed in this appendix.

How a robot moves depends on both the robot and the environment. A six-legged walking robot moves in a way that's different from a two-wheeled balancing robot, which moves differently from a snake-like "slitherbot." However, some aspects of motion are common to all robots, and these are considered first. Next, we probe more deeply into the kinematics of the differential drive base. Because of its simplicity, ruggedness, and reliability, this type of mobility base is one of the most popular. A differential drive base is assumed in most of the examples throughout the text.

Pose

Roboticists refer to the combination of an object's position and orientation as the *pose* of the object. Three numbers are needed to specify the pose of a mobile robot operating on a plane. (It means the same thing to say that a planar robot has three degrees of freedom.) As shown in **Figure A.1**, the three numbers are (x, y, θ), where x and y indicate the position of the center (or any other agreed-on point) of the robot relative in a given coordinate system and θ represents the angle between the x-axis of the coordinate system and the direction the robot is pointing. (Actually, we could use any fixed direction on the robot as long as we use it consistently.)

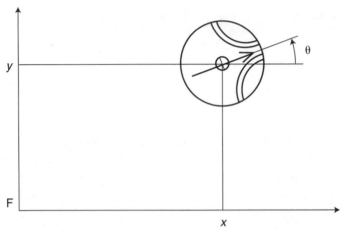

Figure A.1

To analyze robot motion, we first place the robot in a coordinate frame labeled F. The robot has a pose specified by three numbers: (x, y, θ). The center of the robot is located a distance x units from the origin along the horizontal axis (the x-axis) of the coordinate frame and y units above the origin on the vertical axis (the y-axis) of the frame. The forward-pointing axis of the robot makes an angle θ with the x-axis of the coordinate frame.

Velocity is the rate at which position changes. Since three numbers are needed to fix the position and orientation of an object on a plane, a planar object can have three different velocities simultaneously: (v_x, v_y, v_θ). This triplet indicates the component of the robot velocity in the x direction, v_x, the velocity in the y direction, v_y, and v_θ, the rate at which the robot rotates about some point (the robot's center, in this case). But we usually think of the robot as having two types of velocities rather than three. They are a translational velocity, composed of (v_x, v_y), and a rotational velocity, commonly referred to as ω (rather than v_θ).

Dead Reckoning

Access BSim and consult the instructions for running the London task. Make sure to select the operating mode "Fantasy." In this example, the robot follows a path that circumnavigates a collection of obstacles. The robot uses no external sensing to accomplish its task. Given an initial position and orientation, the robot moves forward a certain distance, spins 90 degrees, then repeats the process three more times until it returns to its starting point. Let the simulation run as long as you like; the robot will continue flawlessly following its preprogrammed path.

If we know where the robot starts out (the robot's initial pose) and we know how the robot moves (the three components of the robot's velocity) and we know how long the robot travels at that velocity (the time), then we can figure out where the robot ends up (the robot's final pose). This is the essence of deductive reckoning or, as it is more often called, dead reckoning.

A couple of simple dead-reckoning examples are shown in **Figure A.2**. Suppose our robot begins at point (2, 1, 0); that is, the robot's center is at $x = 2$, $y = 1$, and the robot is pointed outward along the x-axis, $\theta = 0$. Further suppose that the only non-

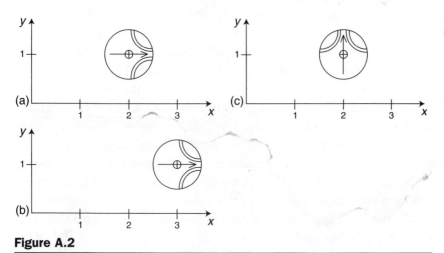

Figure A.2

A robot with velocity in only the x-direction moves from pose (2, 1, 0) in (a) to pose (3, 1, 0) in (b). Or, if the robot starts out with only a rotational velocity, it would end up at pose (2, 1, $\pi/2$), as shown in (c).

zero component of the robot's velocity is in the x direction at one unit per second ($v_x = 1$, $v_y = 0$, $\omega = 0$). After one second, the robot's new position will be (3, 1, 0). The robot will have moved one unit farther out the x-axis, but the y component will not have changed and the robot will still point in the same direction.

Next, imagine that the robot once again starts with pose (2, 1, 0), but that this time the translational velocity is zero (both v_x and v_y are zero) and the rotational velocity, ω, is $\pi/2$ radians (or 90°) per second. In this case, after one second, the robot will be at (2, 1, $\pi/2$). That is, the robot stays at the same position, but ends up pointed parallel to the y-axis.

What happens in the general case when the robot executes both translational and rotational motions simultaneously and when the translational and rotational velocities change with time? The simple, practical way to deal with large, complex robot motions is to decompose them into small, simple pieces of motion that are easy to compute. Adding up all the little pieces of motion gives us the overall motion for the robot. This is known as an *iterative solution.*

An automobile can only go in the direction it is pointed or, more accurately, the direction the wheels are pointed. But a person, with a little effort, can face in one direction and move in another. There is a class of robot mobility bases called a synchro drive that has this same "point in one direction, move in another direction" property. (See the robot illustrated in **Figure 4.12.**) Impressively, we can have such a mobility platform spin like a top while traveling along a straight line! Because the translational and rotational motions of a synchro drive have no effect on each other (we say they are *decoupled*), the motion of such a mobility system is easy to analyze. The programmer can directly command translation, v, and rotation, ω, values. Unfortunately, although the motion of a synchro drive is easy to analyze, the mechanism that implements this ability is complex and thus can be costly to construct.[1]

---

[1]The popular B21 research robot manufactured by the iRobot Corporation incorporates a synchro drive base. Another robot that implements omnidirec-

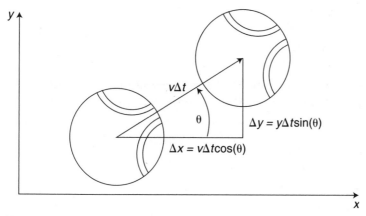

Figure A.3

A differential drive base moves in the direction that it points. If the robot's straight-line velocity is v, then during time Δt the robot moves a distance $v\Delta t$. The x and y components of this motion are $\Delta x = v\Delta t \cos(\theta)$ and $\Delta y = v\Delta t \sin(\theta)$ as indicated. To get the components of the velocity in the two directions, we divide each distance by the time. So $v_x = \Delta x/\Delta t = v \cos(\theta)$ and $v_y = \Delta y/\Delta t = v \sin(\theta)$.

A differential drive base can move only in the direction that it points.[2] This fact places a constraint on the Cartesian component of the velocity (v_x, v_y). As shown in **Figure A.3**, we must have:

$$v_x = v \cos(\theta)$$
$$v_y = v \sin(\theta)$$

where v is the velocity (or more properly speed because here v is not a vector quantity).

The speed of the robot can change in an arbitrary way, as can the robot's rate of rotation. Indeed, these are two of the quantities that a robot program typically controls. But we will analyze robot motion by examining brief time periods where both are nearly constant. So, if we consider the robot's motion over some small time, Δt we have:

---

tional travel by a clever and completely different means is the Palm Pilot Robot Kit developed by Illah Nourbakhsh's Toy Robots Initiative. The kit is available from Acroname, http://www.acroname.com.

[2]There are many different base types and many forms of robot locomotion. For a good review of this topic, see *Robot Mechanisms and Mechanical Devices Illustrated* by Paul Sandin, McGraw-Hill, 2003.

$$x_{n+1} = x_n + v_n \Delta t \cos(\theta)$$
$$y_{n+1} = y_n + v_n \Delta t \sin(\theta)$$
$$\omega_{n+1} = \theta_n + \omega_n \Delta t$$

The position (or orientation) at time $n + 1$ is just the position (or orientation) at time n plus the velocity component (or rotation rate) at time n times the time interval. Δt is a constant of the sys-

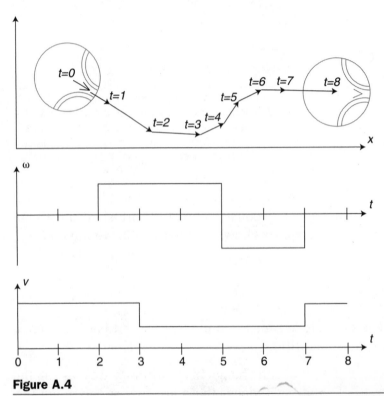

Figure A.4

The sequence of arrows in the top graph illustrates the robot's response to the measured rotation rate, ω, and translation velocity, v, in the lower two graphs. From t_0 to t_1 and t_1 to t_2, ω is zero and v is non-zero; thus the robot moves forward in the direction it pointed at time t_0. The rotation rate is positive between t_2 and t_5 so the robot begins to arc counterclockwise. At t_3 the velocity decreases to half its initial value so the arrows are shorter until t_7 when velocity returns to its initial value. At t_5 the rotation rate becomes negative and the robot arcs clockwise until t_7 when the rotation rate goes to zero and the robot moves straight again. Robot motions are exaggerated for clarity. In practice, the time step, Δt, would be chosen such that step-to-step translations and rotations are very small. Small time steps minimize the error introduced by approximating continuous motion as a series of discrete linear steps.

tem: $\Delta t = tn_1 1 - t_n$. We assume that the robot controller changes the values of v and ω only once every Δt seconds, as indicated in **Figure A.4**.

And that's the dead-reckoning story—to find the robot's position and orientation at any time, start from the robot's known initial position and orientation and repeatedly apply the equations above.

Differential Drive

A differential drive base controls robot motion by moving the two drive wheels differentially; that is, each wheel can move at a specified velocity regardless of the velocity at which the other wheel moves. One or two casters are typically used for balance, as can be seen in **Figure A.5**.

Although translation and rotation of the robot are coupled, we can operationally decouple these components of motion using a

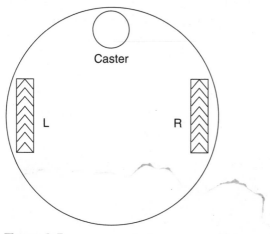

Figure A.5

A popular type of mobility base uses differential drive. Here we look from above, through the robot shell, to see the two drive wheels and the front ball caster (required for balance). The relative velocities of the left and right drive wheels, labeled L and R, respectively, determine robot motion. For example, when both wheels spin forward, the robot moves forward. When the wheels turn in the opposite directions at the same speed, the robot spins in place about a point midway between the two wheels.

two-step process. First, spin the robot in place to point in the proper direction (move the drive wheels in opposite directions at the same velocity), then execute a pure translational motion by turning the drive wheels in the same direction until the robot gets where it is going.

A differential drive base can execute smooth, complex motions. How do the independent velocities of the left and right drive wheels, v_L and v_R, respectively, transform into the translation and rotation velocities, v and ω, analyzed above?

As noted above, we can use any convenient point on the robot as the origin of a local coordinate frame. Calculations are simplified if we choose the point at which the left wheel contacts the ground. This selection is illustrated in **Figure A.6**. Given this choice, the robot's translational velocity, v, is just the velocity of the left wheel, v_L. To find the robot's angular rotation rate, ω, we ask: At what angular rate does the right wheel orbit the left from the point of view of the local frame centered on the left wheel? The velocity, v_{point}, of a point on a circle of radius w, moving at an angular rate of ω, is given by: $v_{point} = \omega w$ or $\omega = v_{point}/w$. Thus, substituting the corresponding values from the robot into the lat-

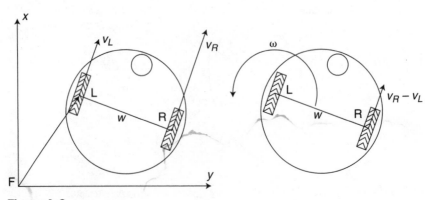

Figure A.6

Relative to a fixed coordinate frame, F, the velocity, v, of a local coordinate system centered on the robot's left wheel is just the velocity of the robot's left drive wheel v_L. The right wheel's velocity relative to the left wheel is $v_R - v_L$. Thus the angular rate, ω, at which the robot rotates is the rate at which the right wheel orbits the left: $\omega = (v_R - v_L)/w$, where w is the distance between drive wheel centers.

ter equation, we can write the relationship between wheel velocities and robot velocity and rotation rate:

$$v = v_L$$
$$\omega = (v_R - v_L)/w$$

The inverse of this solution is also frequently useful:

$$v_L = v$$
$$v_R = \omega w + v$$

It may be helpful to examine a useful special case of differential drive robot motion. In **Figure A.7**, vector r_L is the radius of the circle described by the left wheel as the robot orbits the origin of the coordinate system. As the robot drives around the origin, ϕ goes from 0 to 2π, the right and left wheels describe concentric circles. The circumferences of these circles described by the left and right wheels are $2\pi r_L$ and $2\pi(r_L + w)$, respectively. We can

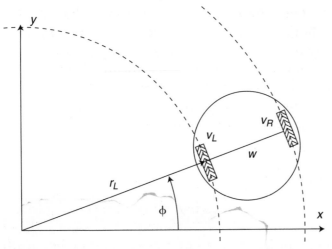

Figure A.7

The diagram shows how wheel velocity relates to robot motion for a differential drive robot with wheel separation, w, when the left and right wheel velocities are held constant. In this case, the robot will orbit about a point. Place the origin of a coordinate system at this point. The paths followed by the robot's wheels are concentric circles (dashed). Although the circumference of each circle is different, the time required to complete each circle is the same (after a complete rotation, the robot must return to the same place facing in the original direction). This fact is used to compute the radius r_L.

compute r_L by noting that the time it takes the left wheel to drive around the inner circle must be equal to the time taken by the right wheel to drive around the outer circle. Since the time is just the distance divided by the velocity, we have:

$$2\pi r_L/v_L = 2\pi(r_L + w)/v_R$$

Solving for r_L we get:

$$r_L = v_L w/(v_R - v_L)$$

This equation, relating the left and right wheel velocities to the robot's orbit radius, has just the right properties. When the two wheel velocities approach the same value (either positive or negative), the radius, r_L, tends toward infinity. A robot moving in a straight line can be thought of as driving around a circle of infinite radius. When the left velocity is zero, r_L is zero and the robot spins about its left wheel. When the left and right wheel velocities have equal magnitudes but opposite signs, the robot spins about its center ($r_L = w/2$). Note that a differential drive robot can be made to rotate around any point along the line connecting the two wheels (including points outside the body of the robot). Note also that negative radii are meaningful in this treatment—a negative radius indicates that the robot arcs to the right; a positive radius implies an arc to the left (arcs are relative to the left wheel).

London Fog

Now that we have the basic idea behind dead reckoning (iteratively adding up small motions to compute the robot's global motion), it is time for some cautions concerning the ugly truth about this method.

The helpful suggestion a roboticist most dreads hearing begins with the words: "Why don't you just…." In suggesting a why-don't-you-just idea, the speaker invariably focuses on solving one aspect of an existing problem while neglecting all the new problems that will be created by adopting the suggestion.

Dead reckoning frequently falls into this category. "Why don't you just give the robot a map, then let it keep track of where it is on the map?" Indeed, now that we know how dead reckoning works, why don't we just do this?

Imagine that you have a map of London and that you want to walk from Piccadilly Circus to Trafalgar Square. Further imagine that a pea-soup fog has descended on the city such that, while you can just make out the map held a few inches before your eyes, your feet are quite invisible in the gloom, as are all sign-posts and street markings. Can you reliably travel from Piccadilly Circus to Trafalgar Square under these conditions?

In theory, yes you can. From the map, you can deduce the distance from one intersection to the next and the number of degrees to turn each time you reach an intersection. Assuming that you know your starting point and initial heading, all you have to do is count your steps and turn precisely when the step count indicates that you have reached an intersection. This is exactly what our example robot does when it executes the London task.

In practice, can you successfully complete your trip? No; your situation is clearly hopeless. Given a map but no possibility of correcting your path along the way, the effects of any tiny mistakes you make early on just grow and grow. After no more than a block or two, you are likely to collide with a building or other obstacle. The collision will inform you that your actual location does not match your putative location on the map, but you will have little clue as to how to resolve the error.

Limits of Dead Reckoning

Run the London task again, but this time select "Noise" rather than "Fantasy" mode. What happens? The program has not changed—the robot is still following the same instruction as before. But somehow the robot doesn't quite return to its original starting position. Over time, the robot strays farther and farther from its proper path and eventually collides with an obstacle. This approximates how mobile robots behave in the real world.

This exercise should reinforce the notion that simulation can be misleading. BSim makes the problem explicit by forcing you to choose to operate either in Fantasy or Noise mode. But many simulators neglect this feature. Had you perfected the London task on a simulator, then run the program on a physical real robot, the robot would not have performed as you expected. Even BSim models only one aspect of the trouble you will encounter running a robot in the real world. Try as we might, we can never eliminate all aspects of fantasy from our simulators.

What are the sources of real-world dead-reckoning noise? One drive wheel is always a little bigger than the other, or not pointed in exactly the same direction as the other, or one wheel always slips a little. But even if you build your drive mechanism with nanometer precision, the environment invariably tosses a monkey wrench into your meticulous preparations. The problem is illustrated in **Figure A.8**. In analyzing robot motion, we implicitly assumed that the robot's wheels are tangent to a common plain and thus the angle through which a wheel turns corresponds directly to the distance traveled in the plain. But if the surface on which the robot moves is not flat, or if there is any dirt or debris on that surface, then this assumption breaks down. In the real world, the assumption always breaks down to the degree that dead reckoning, by itself, is unreliable for anything other than short-range navigation.

Our basic difficulty is that dead reckoning constitutes an open-loop control system. Although we can close the loop on wheel rotation, we cannot close the loop on robot position and orientation; thus, in the absence of some external mechanism, disturbances to those quantities cannot be corrected.

Summary

In this appendix, we have learned about the mathematics of robot motion.

- The motion of a robot can be represented in a coordinate system.

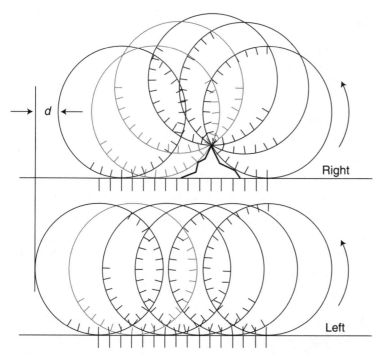

Figure A.8

Unanticipated features in the environment could degrade dead reckoning even if the robot's mechanical system were perfect. Shown here are the starting positions, ending positions, and four intermediate positions of the right and left wheels of a differential drive robot. Both wheels roll from right to left. On flat ground, graduations on the wheels match up with graduations on the ground as the wheels advance. The robot controls the velocity of each wheel such that both turn 150 degrees, as measured by each wheel's shaft encoder. The right wheel, however, encounters a bit of debris over which the wheel must climb; the left wheel rolls on flat floor. In the final frame, both wheels have the same orientation, but not the same position. Although both wheels turn the same number of degrees (the position of each wheel is shown every 30 degrees of rotation), as a result of the encounter with the debris, the right wheel makes less forward progress—the right wheel travels a distance, d, less than the left. This causes the robot to turn slightly to the right.

- A mobile robot moving on a plane has, in general, three degrees of freedom.

- Dead reckoning computes the present position of a robot, given a known starting position by adding up all the small motions the robot makes.

- Dead reckoning is subject to accumulating error.

Exercises

Exercise A.1 How many numbers are required to specify the pose of an object not on a plane, but in a three-dimensional volume?

Exercise A.2 Suppose a synchro-drive robot starts out at (4, 3, $\pi/2$) and wishes to end up at (−2, 1, $\pi/2$) after three seconds. What three components of velocity will achieve this result?

Exercise A.3 Suppose that a differential drive robot moves with constant (scalar) translational velocity and that the left and right wheel velocities are adjusted continually such that the radius of curvature, r, of the robot is $r=a\theta$. (The robot's heading is θ.) What figure does the robot describe as it moves? (Hint, let θ increase starting from zero, but don't reset θ to zero when the value passes 2π.)

Exercise A.4 Carpets exhibit a curious feature that makes dead reckoning especially difficult. The carpet nap preferentially folds over in one direction when weight is applied. The effect of this is to give any object that moves across the carpet a small but persistent nudge in a particular direction. Suppose you have measured the nudge for your carpet such that each time the robot moves one foot north, it moves 0.1 inch east. What heading should the robot take if it wants to move true north? What happens when the robot travels due east or due west? Describe what the robot should do if it wants always to move in a true direction. If the robot's desired x and y velocities are v_{xd} and v_{yd}, write an expression for the velocities the robot should command in order to achieve the desired velocities.

Exercise A.5 In this appendix, we have carefully analyzed the motion of differential drive bases, but said little about the common car-type base (a mobility base using what is known as Ackerman steering). However, we can think of a car-type base as a special case of a differential drive base—one to which we have added certain limitations. Suppose a car-type base can steer the wheels a maximum of 45 degrees from the center position. To simulate this behavior in a differential drive robot, what restriction would we place on the velocity differential between the two wheels? Assume that the steered wheels are separated from each other by a distance w and that the rear drive wheels are w behind the steered wheels. What is the minimum radius arc such a system can achieve? To avoid having any wheel skid while driving in an arc, should both steered wheels turn to the same angle or somewhat different angles?

Exercise A.6   Suppose you build a differential drive mobile robot. You wish to use dead reckoning to compute the location of the robot as it moves about. Each wheel is two inches in diameter. You command the robot to drive straight for 20 feet. How many times must each wheel rotate to make the robot move 20 feet?

Exercise A.7   Now suppose that the left drive wheel from Exercise A.6 has become damaged such that its effective diameter is only 1.9 inches. Assume that the robot's wheels are separated by 6 inches and that the robot starts pointed along the x-axis of a coordinate system. If each wheel rotates the same number of times as before what will be the robot's final location? At what angle, relative to its initial heading, will the robot be pointed? Where will the robot think it is?

Exercise A.8   Using the left wheel as the origin of our local robot as was done in the text may seem less than pleasing. The obvious coordinate system origin is the center of the robot. Make this choice and derive again the relationship between v_L, v_R and ω, v.

B

BSim

BSim is a behavior-based robot simulator designed to allow users to experiment with behavior-based programming techniques without requiring access to an actual robot. (The "B" in BSim stands for behavior.) BSim enables users to create simple worlds of rigid objects and light sources and to program robots to interact with these worlds. Various behaviors and tasks are built into the simulator to give users a feel for what can be accomplished with behavior-based programming. The simulator is accessible as a Java applet on a Web page to permit maximum availability of the software. **Figure 1.1** showed an example of the BSim screen.

BSim is an evolving application. The text below reflects the status of the simulator, as of this writing. However, for the most up-to-date information consult the links you will find at www.behaviorbasedprogramming.com.

BSim Aspects

The Simulator and Time

A discrete time simulator that calls an update procedure on all simulated objects at regular intervals drives BSim. Each simulated object must provide an update procedure for the simulator to call with each time step. Simulated objects can be added directly to the simulator or can be nested in a hierarchical fashion to ensure that specific simulated objects are updated in a predefined order.

The World

A task plus a collection of objects along with a robot or robots in some starting pose is called a simulation. Each simulation defines a simulation object called the World. The World maintains the rigid bodies, lights, and robots in the simulation. When the simulator updates the World, the World first moves all of the rigid bodies according to their velocities, and then allows any robots to sense the World and compute their next actions. The robot sensors have direct access to the World in order to provide sensor readings to the robot.

Elementary physical laws are simulated by the World. The BSim World is two-dimensional and consists only of rigid bodies and lights. The rules governing the movements of rigid bodies are: 1) rigid bodies cannot overlap, 2) rigid bodies can have a translation velocity and a rotational velocity about an arbitrary axis, and 3) a single rigid body can be pushed, but only in straight paths and only by another single, moving rigid body. Lights are not rigid bodies and all rigid bodies are opaque to light. All lights are point sources.

The World performs the physical simulation by first moving all of the rigid bodies according to their translational and rotational velocities. Next, a collision graph is built that links every rigid body with any colliding bodies. Collisions are resolved either by having one rigid body push the other, or, if motion is not allowed, by moving both rigid bodies back to where they were before the simulated time step. If no rigid bodies were in collision at the start of the time step, then we can be guaranteed that no objects will be colliding at the end of the time step by simply moving the relevant objects back to their previous positions if an object cannot be pushed.

All rigid bodies are constrained to be in the form of convex polygons. With convex polygons, collisions are easily detected by searching for a separating plane between two rigid bodies. To determine the visibility of a light source, the World checks to see if a line from a light source to the photo sensor intersects any rigid bodies in the world. Rigid bodies can be made into composite objects by attaching objects to the rigid body. Objects can only be attached to a rigid body if they define a procedure that the rigid body can use to set that object's position. Photo sensors are added to robots using this technique.

Fantasy Mode and Noise

By default, the simulator is run in Fantasy Mode, which means that no noise is added to the robot's sensors and effectors. By switching Fantasy Mode off, random noise is added to the robot's wheel speeds. Sensor values are unaffected by noise. The Fantasy Mode flag is stored in the World object.

Latency

Sensors simulated in BSim can exhibit latency by setting the World's latency parameter. The latency parameter specifies how many time steps pass before the current sensor readings are available from the sensor. Each sensor keeps a history of its measurement values at least as far back as the latency parameter requires. If a behavior requests a sensor value before the value is available, the sensor signals to the behavior that it must wait.

A Simple Robot

BSim implements a simple yet functional simulated robot (depicted in **Figure 1.2**) with a differential drive mobility base and multiple sensors. The BSim robot is round in shape and has two shaft-encoded drive wheels on either side of the robot's body. The round body shape enables the robot to spin regardless of the configuration of adjacent obstacles. Two proximity sensors and two photo sensors are mounted on the front of the robot on either side facing diagonally outward. Each robot can be configured to support a set of behaviors. The user programs a robot by selecting a set of behaviors from a built-in list of predefined behaviors. The user selects the relative priorities of these behaviors and sets the parameters governing the functionality of each behavior. Robots use fixed-priority arbitration.

The robot provides procedures that allow access to the left and right wheel speeds, the left and right photo sensor values, the state of the proximity sensors, the inputs from a remote control, and whether the robot is bumping or colliding with another rigid body. Finally, the robot provides a procedure for reporting the behavior that was run by the arbiter during the last time step.

Behaviors and Arbitration

BSim includes a set of predefined behaviors to facilitate the rapid construction of interesting robot tasks. Each behavior is referenced to an arbiter and each behavior defines an action procedure that is called at each time step. The arbiter provides each behavior with a stub interface to the robot; the behaviors use this to request access to the robot's sensors and effectors. In the action procedure, the behavior looks at the robot's sensor values and requests that the robot perform specific actions by making calls to the arbiter's stub implementation. The behaviors run continuously and the arbiter chooses a winner from among the behavior requests.

Behaviors maintain state and can be configured with a set of parameters. Some behaviors require access to time information, so behaviors also provide an update procedure for the simulator to call. Because rigid bodies in the World can be pushed only in straight paths, behaviors drive the robot in curved paths by first driving the robot forward and then pivoting. In this way, the robot can appear to push objects along curved trajectories. The parameters that a behavior supports are discovered at run time using reflection; this allows the easy addition of new behaviors to BSim.

Cruise

The Cruise behavior is the simplest of the predefined behaviors. It has two parameters, the left and right wheel speeds. With each time step, the Cruise behavior requests that the left and right wheel speeds be set to the prescribed values.

Escape

The Escape behavior is a ballistic behavior performed in three steps: backup, spin, and forward. It takes three parameters—the lengths of time that Escape should perform each of its three steps, t_{backup}, t_{spin}, and $t_{forward}$, respectively. The behavior is implemented as an FSM with one state for each step in the behavior, such that the requests made by the behavior depend on the behavior's state. Each state checks how much time has passed since the behavior entered that state. After the prescribed time for that state has passed, the behavior transitions to the next state.

The behavior begins in a START state. START does nothing until the robot collides with an obstacle. The behavior then transitions to the BACKUP state. BACKUP drives the robot backwards until the time t_{backup} has passed. When BACKUP completes, the behavior enters either the SPIN_LEFT or SPIN_RIGHT state, depending on a random coin flip. (The BSim bumper knows only that a collision has occurred; it does not know on which side of the robot the collision happened.) Both spin states rotate the robot in place in one direction or the other for a time t_{spin}. After completing the spin, the Escape behavior enters the FORWARD state. Here the robot moves the robot forward for a time $t_{forward}$, then the Escape behavior transitions back to the START state. (For more information, see the section, "FSM Implementation," in Chapter 3.)

Avoid

The Avoid behavior (given a positive gain parameter, g) moves the robot forward and left if the right proximity sensor is on, or forward and right if the left proximity sensor is on. This tends to make the robot avoid obstacles. If g is negative, the robot turns toward obstacles. The magnitude of g determines how tightly the robot arcs.

Wall Follow

The robot can use its left and right proximity sensors to perform wall following. In a fashion similar to the Escape behavior, Wall Follow begins in a START state, remaining there until either the left or right proximity sensor detects an object. Whenever the robot detects an object on the left, the behavior enters a LEFT state that moves the robot forward and to the right. If the robot detects an object on the right, the behavior enters a RIGHT state. In this state, the robot moves forward and to the left.

From the LEFT and RIGHT states, once the robot has veered so far away that its proximity sensors are not detecting anything, the robot enters a LEFT_LOST or RIGHT_LOST state. The LEFT_LOST state moves the robot forward and left in an effort to search for the wall and the RIGHT_LOST state moves the robot forward and right. This continues until either a proximity sensor is triggered (transitioning the behavior back to the LEFT or RIGHT state), or the lost states time out and the behavior goes back to the START state. If at any time the last chosen behavior was not the wall-following behavior, the behavior immediately resets to the START state.

Home

The Home behavior tries to drive the robot toward a light source. Home uses a proportional controller to home on a light source whenever the robot's photo sensors see light. The robot homes on the light by pivoting in the direction of the light and then moving forward a step. The robot determines the direction to the light by calculating the difference between the two photo sensor measurements. A gain parameter determines how rapidly the robot turns toward the light, and a speed parameter determines how fast the robot moves.

Anti-Moth

The behavior Anti-moth is typically inactive most of the time. This behavior triggers whenever the total light intensity measured by the robot's photocells exceeds a parameter value chosen by the user. When Anti-moth triggers, the robot is forced to spin left or right for a random amount of time. Anti-moth thus keeps the robot from getting too near the light.

Dark-Push

The Dark-push behavior resembles the Escape behavior when the robot bumps something in the dark. If the robot encounters a pushable object when no light is visible, Dark-push triggers. This behavior prevents the robot from pushing rigid bodies away from the light source toward walls. Dark-push is implemented using an FSM similar to the one used by Escape, but with a different trigger. Dark-push is triggered by no light plus an object-is-being-pushed indication from the bumper.

London

The London behavior moves the robot in a rectangular path. An FSM implements this behavior. London is able to perform precise 90-degree turns because its implementation is carefully tuned to the timing of the simulator. However, if the simulator operates in Noise rather than Fantasy mode, the London behavior will gradually wander away from the desired path.

Gizmo

The Gizmo behavior adjusts the robot's position relative to a light source such that the sensed brightness matches a preselected value. If the light is too dim, the robot drives forward, and if the light is too bright, the robot drives backward. The behavior drives the robot in a straight path and sets the wheel speeds using a proportional controller. Gizmo takes as parameters the gain for the controller and the target brightness.

Remote

The Remote behavior responds to commands from a user-controllable joystick panel on the BSim window. This control can drive the robot forward, backward, and pivot the robot left and right. The robot interface includes the remote signals so that the behaviors can access them as sensory inputs.

Tasks

A task is a set of parameterized behaviors combined with an arbiter; a task performs a specific function. BSim includes a number of predefined tasks described below. Some tasks have the same name as their principal behavior.

Collection Task

The Collection task brings scattered pucks to a light source. As described earlier (see **Figure 1.3**), the Collection task uses behaviors Escape, Dark-push, Anti-moth, Avoid, Home, and Cruise in order of decreasing priority. Under the control of Cruise, the robot drives straight until it comes to an object. If the IR sensors detect the object, the robot turns toward the object. Upon colliding with the object, one of two things occurs. If the object is an immovable wall, then Escape takes control and drives the robot away from the object. If the object is a movable puck, then the robot begins to push the puck. If the robot is pointed away from the light source, Dark-push becomes active and forces the robot to turn away—abandoning the puck. If a light source is visible to the robot, then Home becomes active and the robot homes on the light. When the robot approaches too close to the light and the apparent brightness becomes too great, the Anti-moth behavior triggers turning the robot away from the light.

Gizmo Task

The Gizmo task loads the Gizmo behavior. This actively positions the robot a set distance from a light source. The task also includes the Escape behavior for robustness.

London Task

The London task loads the London behavior, which drives the robot in a square path. The Escape behavior is not added here, in order to show how open loop control can fail, allowing the robot to wander from its path and become stuck.

Simulations

BSim can load predefined simulations. Each simulation specifies a World, the value of the latency variable, and the setting of the Fantasy/Noise mode.

Collection Simulation

The Collection Simulation demonstrates the Collection Task. A robot is positioned in a room with scattered pucks and a central light source. As the simulation runs, the robot moves the pucks to the vicinity of the light source.

London Simulation

The London Simulation demonstrates a weakness of open loop control. A robot programmed with the London Task is placed in a square-shaped corridor. The robot follows the corridor perfectly in Fantasy Mode, but will eventually get stuck once Fantasy Mode is switched off and noise is added to the system.

Gizmo Simulation

The Gizmo Simulation demonstrates proportional control using the Gizmo task. A robot programmed with the Gizmo Task is positioned facing a light source. Initially, the gain for the Gizmo behavior is set to a low value, while World latency is set to a high value. As the simulation runs, the robot moves slowly toward the light source, stopping when a specified light intensity is reached. If the gain is increased and the simulation restarted, the robot will oscillate around the point of desired light intensity.

User Interface

World Editor

Simulated objects can be added to the world by selecting the object you wish to add from the tool panel and then clicking on the world panel. To add pucks, select the puck tool, then click on the world panel at the place where you want the puck to appear. Robots and lights can be added in a similar fashion. To add walls, click on two points to designate either end of the wall. BSim does not allow you to add rigid bodies that overlap, so use care in placing objects. Predefined simulations can be loaded from the BSim graphical user interface by selecting the Simulations menu. Fantasy Mode is turned on and off by selecting Fantasy Mode from the Options menu. Selecting Set Latency from the Options menu allows the value system latency to be adjusted.

The Robot Programmer

BSim includes a graphical user interface for programming a robot. To access the robot's program, double click on a robot. A dialog box displaying two lists appears. The behaviors list (on the left) indicates all the behaviors the robot supports. The task list on the right displays the behaviors included in the robot's current task in order of decreasing priority, starting at the top. Behaviors can be added or removed from the task list and the order of the list can be changed. Selecting a behavior in the task list brings up a property sheet for that behavior. The property sheet provides a way to modify the behavior's parameters. Predefined tasks can be loaded into the task list by selecting a task from the Tasks menu.

To add a behavior to the current task, either double click the behavior in the behaviors list, or select the behavior you wish to add and click the Add button. To remove a behavior, double click the behavior in the task list, or select the behavior and click the Remove button.

To change the value of a parameter for a behavior, select the behavior in the task list. The behavior's parameters will then be displayed in the property sheet next to the task list. Click on the value you wish to change. The value should now appear in a white text box. Type in the new value and press Enter. The value should appear gray again.

To load the task in the task list into the robot, click the Load button. To cancel and leave the robot unchanged, click the Cancel button.

C

Frequently Used Functions

In the code for the various robotic systems I have developed over the years, a small number of simple functions tend to appear again and again. Although far from profound, the functions below are often helpful.

Clip

The result of an arbitrary computation can be a number of any size, but the number that a microprocessor can store in a variable and the numeric value that can be meaningfully sent to an actuator have definite limits. Thus it is frequently necessary to restrict the value returned by a computation to a manageable range. A very simple function of three arguments that can do this is clip.

```
Function clip(value, minimum, maximum)
    If (value < minimum)
        Return minimum
    else if (value > maximum)
        Return maximum
    else
        Return value
```

```
      end if
end clip
```

Clip chops off values that are larger than the maximum or smaller than the minimum allowed.

Leaky Integrator

The leaky integrator is a helpful technique to use any time you need a limited amount of state, that is to say, when you want your robot to remember something, but only for a while.

We can use a leaky integrator to implement a version of the anti-canyoning behavior we saw in the section, "Anti-Canyoning," in Chapter 5. See also **Figure C.1**. In the anti-

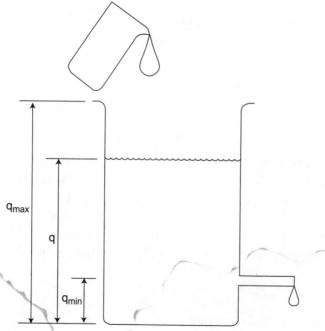

Figure C.1

Leaky integration is the mathematical analog of pouring water into a bucket with a hole in it. The quantity, q, being integrated (or summed up) is increased by some series of external events. There is a maximum value of q, q_{max}, analogous to the height of the bucket and a minimum value of q, q_{min}, analogous to the height of the hole. The external events add water to the integrator while the built-in hole allows the water to leak slowly away.

canyoning example, we want the occasional detection of an obstacle to increase the number of degrees the robot spins when it next encounters an obstacle. But we also want the turn-angle to decay back to zero after the robot fails to see an object for a while. The following initializations and code can accomplish this.

```
Parameters
    check_interval = t₀        //Wait this long between updates of the
                               // integrator
    leak = θ₀                  //This many degrees leak away each update
Variables
    leaky_sample               //Anti-canyon sets sample, leaky_integrator
                               // clear it
    leaky_sum                  //The "water level"

Process leaky_integrator (decay, min, max)
    leaky_sum = clip (leaky_sum + leaky_sample - decay, min, max)
    leaky_sample = 0           //Reset sample
end leaky_integrator

Start_process(check_interval, leaky_integrator(leak, min_turn, 180))
```

To implement an anti-canyoning behavior, we set up the leaky_integrator process. The leaky integrator maintains a value that the robot interprets as the angle to turn upon encountering an obstacle. Every time the robot switches its idea of which side the obstacle is on, we arrange for the leaky_sample variable to become non-zero—that is, we add some "water" to the integrator. The leaky_integrator process's argument, min, is min_turn (the robot tries to avoid the object by turning this small number of degrees at first), max is 180 degrees (it makes no sense for the robot to spin farther than is needed to point in the opposite direction), and decay is the number of degrees that should leak away, the variable leak, each time the process runs. The parameter check_interval, invoked when the process is started, determines how often the leaky integrator code runs. For best operation, we must choose a sufficiently small number for check_interval such that Anti-canyon does not change leaky_sample more often than the leaky_integrator process runs.

The result of setting up the code in this way is that when the robot detects a wall, first on one side and then the other, leaky_sum gets bigger. Leaky_sum controls how far the robot turns when it tries to avoid an object, thus the robot swings through larger and larger arcs. After the robot has traveled for awhile without detecting an obstacle, leaky_sum dwindles back to min_turn. This means that the robot will react weakly upon detecting an object after not having detected an object for a long while—the robot forgets about its earlier encounters.

Running Average

Computing the running average of some quantity can be very helpful for mitigating the effects of noise. Suppose your robot needs to act on a signal that is affected by a good deal of noise. Each time the robot reads the sensor supplying the signal, it takes a new sample, S_n of the signal. The first reading of the signal is S_1, the 100th reading is S_{100}, and so on.

If the noise is statistically random, then the noise increases the measured value of the signal as often as it decreases the measured signal. Thus, if you take an average of the signal over some number of samples, the noise will average to zero while the signal will average to its true value. (We assume the signal varies slowly relative to the frequently at which samples are taken.) The question to ask then is: Over how many samples should you average? Also: When should you start and when should you stop averaging? The answer to the first question depends on the nature of the signal and the noise, and can only be addressed by thinking about your particular system. But the second question yields to a bit more analysis. If you decide that you need to average over 10 samples, then one method would be arbitrarily to pick a time to start, take a new sample every time the clock ticks, and 10 ticks later compute the average. But because you get a new answer only once every 10 ticks, your program doesn't get the benefit of the most up-to-date information.

A better method is to keep a running average over some number of recent samples, as illustrated in **Figure C.2**. One frequently used formula for computing such an average is:

$$\bar{S}_n = \frac{1}{k}S_n + (1 - \frac{1}{k})\bar{S}_{n-1}$$

Here $\bar{S}_n$ is the average after n samples have been taken and S_n is the nth sample. The number k (with a value of one or greater) relates to the number of samples that influence the average. It is easy to see that if k is 1, then only the most recent sample contributes to the average—the average changes as rapidly as the sample. But if k is, say, 100, then only 1 percent of the value of the average is due to the most recent sample; 99 percent of the value derives from earlier samples. Thus the average changes slowly.

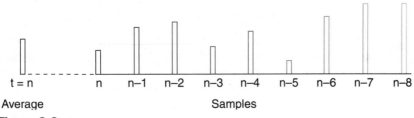

Figure C.2

A running average weights recent samples more heavily than samples from farther in the past.

Angle Computations

In calculating where to go and how far to turn, robots necessarily perform computations involving angles. But angles are awkward computational objects because, unlike regular numbers, angles wrap around. Add a couple of large angles and you can end up with a small angle. For example, the sum of 180 degrees and 190 degrees is often most usefully thought of as 10 degrees rather than 370 degrees. Also, it is not always clear what to do with the sign of an angle. For example, is –20 degrees always the same as 340 degrees?

Part of the confusion stems from the practice of using angles in two distinctly different ways. An angle (along with a radius) can be used 1) to specify a point in a coordinate system or 2) to indicate a rotation. This is illustrated in **Figure C.3**. The convention that I find useful is to consider angle coordinates as positive numbers only in the range of 0 to 360 degrees (or 0 to 2π radians). When used as a rotation, an angle should have values between –180 and 180 degrees (or $-\pi$ and π radians). This range is appropriate because a robot can rotate in the positive or negative direction, but it rarely makes sense to rotate farther than 180

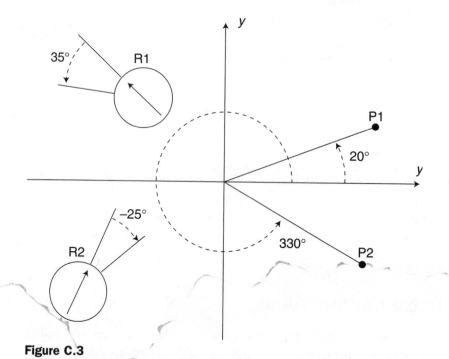

Figure C.3

Angles can represent coordinates or rotations. Computational confusion is minimized if angular coordinates occupy the range 0 to 360 degrees and rotations are allowed to range from –180 to 180 degrees. Here point P1 has an angular coordinate of 20 degrees while point P2 has an angular coordinate of 330 degrees. Robot R1 makes a positive rotation of 35 degrees and robot R2 rotates 25 degrees in the negative (clockwise) direction.

degrees. If, for example, a robot responds to an obstacle blocking its path by rotating farther than 180 degrees, then the robot could have saved time by choosing to rotate a smaller number of degrees in the other direction.

Two functions can help keep angle computations normalized to the proper range. When the result of an operation is to be interpreted as a coordinate, apply norm_360 to the result; when a rotation is the end product, use norm_180.

```
Function norm_360 (raw_angle)
   Return  mod(raw_angle, 360)
end norm_360

Function norm_180 (raw_angle)
   Return (norm_360(raw_angle + 180) – 180)
end norm_180
```

Both functions make use of the usual mathematical definition of the mod operator. (Mod is equivalent to the C operator % for positive integer arguments.) A couple of examples illustrate the point. If a robot at angular coordinate 10 degrees executes a rotation of –30 degrees, the sum of coordinate and rotation is –20 degrees. To normalize, we would use norm_360(–20) giving the resulting angular coordinate of 340 degrees. Or, to discover the rotation needed to change heading from 355 degrees to 5 degrees, we would subtract 355 from 5 to get –350 degrees and use norm_180(–350) to compute a rotation of 10 degrees.

D

Pseudocode

Most behaviors and other programming constructs described in the text are implemented in pseudocode. Pseudocode resembles a computer program in structure, but need not be a legal statement in any particular computer language. Pseudocode is useful because it helps us understand how a program might be written while leaving out the distracting and often idiosyncratic details of any particular language.

As an example, here are a few lines of pseudocode that implement a dead zone and saturation function described in the section, "Saturation, Backlash, and Dead Zones," in Chapter 2.

```
Function xfer (signal, saturation, dz_half_width)
   If |signal| < dz_half_width then
      Return 0
   else if |signal| > saturation then
      Return signum(signal) * saturation
   else
      Return signal
   end if
end xfer
```

The first line creates a function (you can think of a function as a subroutine) named xfer. Xfer takes three arguments, signal—the quantity whose value we wish to limit; saturation—the maximum magnitude that signal can have; and dz_half_width—half the width of the dead zone. The following lines of code examine the value of signal and return an appropriate number. When the absolute value of signal is less than dz_half_width, the function xfer returns 0—signal is within the dead zone. The Return instruction terminates the function. Executing the Return statement transfers control back to whatever bit of code called the function. When the absolute value of signal is greater than saturation, xfer returns positive or negative saturation. (Signum is a function that returns +1 if its argument is greater than zero; it returns −1 if the argument is less than zero.) Otherwise, the signal is neither saturated nor within the dead zone, so xfer returns signal.

All functions expressed in pseudocode follow this basic format: the word Function, followed by the name of the function, followed by any arguments in parentheses. Subsequent lines contain body code. The last statement is always the word end followed by the function name. Comments, introduced by "//" can be included in pseudocode.

Behaviors expressed in pseudocode follow a similar format. The first line is the word "Behavior" followed by the name of the behavior. Unlike functions that can be called by any other code including other functions, only the scheduler calls behaviors.

Several standard variable types are implicitly associated with a behavior. One such type is the parameter. Parameters, as explained earlier, are variables that the programmer uses to specialize the operation of a behavior. A behavior may also have local variables. These variables exist for the convenience of the programmer. Typically, the value of a local variable is lost when control is transferred from a behavior back to the scheduler. Persistent (or static) variables are variables whose values are maintained between executions of a behavior. Behaviors can have inputs; often, sensors generate these inputs. Behavior outputs are usually either robot motion commands (rotation and translation velocities) or motor commands (left and right wheel velocities). Another type of output that a behavior may be required to have is an active flag or triggered flag. This is a logical value that is true when the behavior wants to gain control of the actuators; it is set to false when the behavior is content to command no particular action. A behavior's body code processes sensory information, decides if the behavior wants control, and computes commands for the robot's actuators. Fully specified, a behavior has a form analogous to this example:

```
Behavior Behavior_name
    Parameters: Par_1, Par_2, ...
    Local: Var_1, Var_2, ...
    Persistent: Static_1, Static_2, ...
    Inputs: Sensor_1, Sensor_2, ...
    Outputs: Translation, Rotation, Active, ...
    <body code>
end Behavior_name
```

Generally, we use a simplified version of this formalism. We understand that the robot's software system calls each behavior in turn, giving each a chance to run as often as possible. The scheduler in a real robot might give each of its behaviors a chance to run tens or hundreds of times per second.

A process is a block of code that is run at regular intervals by the robot's software system. The format of a process is analogous to that of a function or behavior. The code that starts a process (a start_process function) specifies how frequently the process will run.

E
Bibliography

Arkin R. *Behavior-Based Robotics.* Cambridge, MA, MIT Press, 1998.

Arkin R. "Motor Schema Based Navigation for a Mobile Robot: An Approach to Programming by Behavior," *Proceedings of the IEEE Conference on Robotics and Automation.* Raleigh, NC, 1987, 265–271.

Braitenberg V. *Vehicles, Experiments in Synthetic Psychology.* Cambridge, MA, MIT Press, 1984.

Brooks RA. "A Robust Layered Control System for a Mobile Robot," *IEEE Journal of Robotics and Automation.* RA-2, 14–23 April, 1986, 14–23.

Brooks RA. *The Behavior Language; User's Guide,* A.I. Memo 1227. Cambridge, MA, MIT AI Lab, April 1990.

Connell J. *A Colony Architecture for an Artificial Creature,* AI-TR 1151. Cambridge, MA, MIT Press, 1989.

Doty K, Harrison R. "Sweep Strategies for a Sensory-driven, Behavior-based Vacuum Cleaning Agent," *AAAI 1993 Fall Symposium Series, Instantiating Real-World Agents.* Raleigh, NC.

Everett HR. *Sensors for Mobile Robots,* Natick, MA, A K Peters, Ltd., 1995.

Franklin G, Powell J, Emami-Naeini A. *Feedback Control of Dynamic Systems.* 3rd ed, Addison Wesley, 1994.

Gage D. "Randomized Search Strategies with Imperfect Sensors," *Proceedings of SPIE Mobile Robots VIII,* Sept. 1993, Vol. 2058, 270–279.

Horswill I. (course notes) http://www.cs.northwestern.edu/academics/courses/special_topics/395-robotics/.

Jones J, Flynn A, Seiger B. *Mobile Robots: Inspiration to Implementation.* Natick, MA, A K Peters, Ltd., 1999.

Jones J, Wirz B. "RoCK: the Goal of a New Machine," *Circuit Cellar Magazine*, April–July, 2002.

Lozano-Pérez T, Jones J, Mazer E, O'Donnell P. *Handey: A Robot Task Planner.* Cambridge, MA, MIT Press, 1992.

Lunt K. *Build Your Own Robot*, Natick, MA, A K Peters, Ltd., 2000.

Martin F. *Robotic Explorations*, Prentice-Hall, Inc., 2001.

Moravec H. *Mind Children.* Cambridge, MA, Harvard University Press 1998.

Moravec H. *Robot: Mere Machine to Transcendent Mind.* Oxford University Press, 1999.

Nisley E. "Above The Ground Plane: IR Sensing," *Circuit Cellar Magazine*, August 2003, 40–43.

Paul RP. *Robot Manipulators: Mathematics, Programming, and Control.* Cambridge, MA, MIT Press 1982.

Raucci R. *Personal Robotics: Real Robots to Construct, Program, and Explore the World.* Natick, MA, AK Peters, Ltd., 1999.

Sandin P. *Robot Mechanisms and Mechanical Devices Illustrated.* New York, McGraw-Hill, 2003.

Stubberud A, Williams I, DiStefano J. *Schaum's Outline of Feedback and Control Systems.* New York, McGraw-Hill, 1994.

Thrun S, et al. "Probabilistic Algorithms and the Interactive Museum Tour-Guide Robot Minerva," *International Journal of Robotics Research.* Vol. 19, No. 11, 972–999.

Turing A. "On Computable Numbers, with an Application to the Entscheidungsproblem," *Proc. London Math. Soc.* Ser. 2–42 (Nov. 17, 1936), 230–265.

Walter WG. *The Living Brain.* New York, W. W. Norton & Company, Inc. 1953.

Weizenbaum J. *Computer Power and Human Reason.* Cambridge, MA, MIT Press, 1976.

Index

Note: Boldface numbers indicate illustrations; "f" indicates footnote.

About the Authors

Joseph L. Jones is a veteran roboticist, currently employed by iRobot. He is the co-author of *Mobile Robots: Inspiration to Implementation*, as well as numerous technical articles in the field of robotics. He is one of the inventors of the Roomba® Robotic FloorVac, the first widely adopted consumer robot.

Daniel Roth is a master's degree candidate in computer science at the Massachusetts Institute of Technology, specializing in autonomous robot navigation.